V. R. Solanki
Vasudha Lingampally
Sabita Raja Sangam

Limnology Of Bellal Lake, Bodhan, India

V. R. Solanki
Vasudha Lingampally
Sabita Raja Sangam

Limnology Of Bellal Lake, Bodhan, India

LAP LAMBERT Academic Publishing

Impressum / Imprint

Bibliografische Information der Deutschen Nationalbibliothek: Die Deutsche Nationalbibliothek verzeichnet diese Publikation in der Deutschen Nationalbibliografie; detaillierte bibliografische Daten sind im Internet über http://dnb.d-nb.de abrufbar.
Alle in diesem Buch genannten Marken und Produktnamen unterliegen warenzeichen-, marken- oder patentrechtlichem Schutz bzw. sind Warenzeichen oder eingetragene Warenzeichen der jeweiligen Inhaber. Die Wiedergabe von Marken, Produktnamen, Gebrauchsnamen, Handelsnamen, Warenbezeichnungen u.s.w. in diesem Werk berechtigt auch ohne besondere Kennzeichnung nicht zu der Annahme, dass solche Namen im Sinne der Warenzeichen- und Markenschutzgesetzgebung als frei zu betrachten wären und daher von jedermann benutzt werden dürften.

Bibliographic information published by the Deutsche Nationalbibliothek: The Deutsche Nationalbibliothek lists this publication in the Deutsche Nationalbibliografie; detailed bibliographic data are available in the Internet at http://dnb.d-nb.de.
Any brand names and product names mentioned in this book are subject to trademark, brand or patent protection and are trademarks or registered trademarks of their respective holders. The use of brand names, product names, common names, trade names, product descriptions etc. even without a particular marking in this works is in no way to be construed to mean that such names may be regarded as unrestricted in respect of trademark and brand protection legislation and could thus be used by anyone.

Coverbild / Cover image: www.ingimage.com

Verlag / Publisher:
LAP LAMBERT Academic Publishing
ist ein Imprint der / is a trademark of
OmniScriptum GmbH & Co. KG
Heinrich-Böcking-Str. 6-8, 66121 Saarbrücken, Deutschland / Germany
Email: info@lap-publishing.com

Herstellung: siehe letzte Seite /
Printed at: see last page
ISBN: 978-3-659-45710-4

Table of Contents

Introduction

Water works for us in many ways, making our lives easier and more enjoyable. But we must take great care not to overuse and abuse this precious resource.

Earth is the only known planet with water. Humans, who can last a month without food, die after a week without water. Most of us are not aware of the vast network of reservoirs, plants and pipes -- infrastructure -- that provides processes and treats our water.

Water is one of the important commodities which man has exploited than any other resource for the sustainance of his life. Over 97% of the water on this earth is stored in oceans and ice caps, which is difficult to be recovered for our diverse needs. Water is a renewable resource. Water is the most basic and most important of all natural resources. Just over 3% of the total water is involved in the global cycle and most of that is locked up in the polar ice caps and in glaciers. We require regular supply of water, but it must be fresh water from rivers, lakes, and underground aquifers.

Lucas a pioneer water biologist has described pure water. Pure water is colourless, odorless, tasteless, easy to flow and devoid of turbidity. These are the five characteristics which make the water pure and clean. Pure water rarely occurs in nature because of its capacity to dissolve numerous substances in large quantities. Water obtained from precipitation is derived from condensation of water vapor in the atmosphere. It is practically only the distilled water which is an exception of impurities from atmosphere. It is only the earth's crust that obtains a number of in organic and organic salts and suspended materials that may impact colour, taste, odor, hardness, alkalinity, acidity and other chemical characteristics to natural water. Most natural waters are therefore, weak solutions of various salts, gases and organic compounds. When these are limited or in reasonable limits it is harmless to the living organisms, but when they exceed, they become hazardous to the nature and biological fauna and flora.

Water is a basic necessity of life, not only for people but for all plants and animals. Water accounts for about 65% of our body weight. If we lose as little as 12% of it, we would soon die.

Water is essential not only for survival but also contributes immeasurably to the quality of our lives. Since the dawn of time, human beings have harnessed water to improve their lives. In some ways, the history of civilization is the story of how we have made water work for us in ever more ingenious ways. As early as 5000 B.C., our predecessors used irrigation to increase crop production. Archaeologists have found masonry sewers dating back to 2750 B.C. and water-flushed toilets dating back almost as far.

Water transportation is still the most efficient way to move bulk goods. Water is also the basis of cheap energy. It is a raw material in the manufacture of chemicals, drugs, beverages, and hundreds of other products. It is an essential part of the manufacturing processes that produce everything from airplanes to zippers. In other words, we depend on water for most of our technology, comforts and conveniences, and of course for personal hygiene and to flush away our waste products.

Many people think it makes no difference how much water we use or what we use it for. Actually, the way we use water is very important. Some uses are incompatible with others. Some uses remove water from the natural cycle for longer periods than others. Worst of all, most uses actually lower the quality of the water.

Water quality is everybody's business because ultimately we all draw from the same supply of water. We are now aware of limits to the reuse of water, when and where it is returned to nature diminished in quantity and quality. Therefore, we must learn to understand water use much better: where we use it, what to measure, what the main uses are, how they compete and interfere with each other, and how to manage the growing competition.

In India the annual rainfall is approximately 105 cm (Sharma, 1987) of the total rainfall, which falls on the earth 36.75 cms rainfall evaporates from the surface; 42 cms flow down to sea as runoff water and 26.25 cms is stored in the pores of soil

(Singh, 1985). Out of 105 cms rainfall a small fraction is left as surface water present in ponds, lakes, springs and swamps Brookes, streams creeks and rivers which contribute our main water resources.

India has most varied inland water resources, which are considered to be one of the richest in the world, natural lakes; however they are largely confined to the Himalayan region while the rest of sub-continent is dominated by river systems and manmade impoundments. Although global demand for water continues to rise, the development of water resources in some countries has been restricted by regional water storages and environmental damage associated with existing water resource schemes. Complex systems have been developed to try, to match the demand and supply; overtime and from place to place. Water storage is major and recurrent problem in many countries including India.

In comparison to oceans and terrestrial, inland fresh water bodies constitute one-fifth of the earth's surface. Lakes are short lined features of the earth crust and because of its limited dimensions are associated with a different set of problems. The epilimnion of impounded waters, constituting a part of natural habitat, has its own physical and chemical characteristics and holds large number of organisms (Hussainy 1966). The impounded water is considered to be more closed type of ecosystem and any deteriorations of this natural system lead to rapid deterioration of such an ecosystem.

Lakes have been used as ideal natural laboratories to study a number of processes that are important in understanding hydro biological processes including evaporation, dissolution mixing, precipitation of minerals and chemical exchange between water and atmosphere, and diversity of living organisms like planktons, flora and fauna and microbial diversity.

The quality of life is linked with the quality of environment; hence the biological components of fresh water depend solely on latter's physico-chemical conditions. Analysis of physico-chemical parameters of water is therefore essential, as it has great bearing on the explanation of the aquatic ecosystem. The planktons are the indicators of physicochemical and ecological conditions and reveals past history of

5

lake, and the chemical nature shows its recent conditions.The fact that lakes occupy such a small fraction of the landscape belies their importance as environmental systems and resources for human use. They have intrinsic ecological and environmental values. Besides, human use, they are used for many commercial purposes including fishing, transportation, irrigation and industrial water supply, and function as receiving waters for wastewater effluents.

The nutrient enrichment leads to undesirable change in the structure and function of an ecosystem (Smith et al. 1999). Eutrophication is a potent threat to the biodiversity of aquatic environment .Anthropogenic nutrient enrichment causes serious alterations in the coastal zone ecosystem.

The utility of planktons as direct or indirect food for fishes and their utility in assessing the water   quality have now been well established. Water pollution is characterized by the deterioration of the quality of land water or marine water as a result of human activities nearby .The related human activities such as mining, agriculture, stock-breeding, fisheries, urban human activities leather industry, forestry, construction work and various other industries. If there is any change in the physico-chemical characteristics that are adversely affect the living beings. This is worldwide problem both for industrialized and developing countries, as well as in the rich and poor countries.

Eutrophication has been proved to be one of the most widespread and serious anthropogenic disturbances to aquatic ecosystems. The major cause for eutrophication is increased loading of nutrients, especially phosphates (PO_4). Increasing wastewaters, introduction of phosphorous containing detergents, use of fertilisers, and erosion in the watershed are the major reasons for increased loading of nutrients. Fertilisers are the major source of nitrogen, while phosphorous in the fertiliser may be immobilised in the soil. Flooding can however transport large quantities of phosphorous due to leaching from the soil.

Water quality alterations results in disturbances to biodiversity: the fresh waters of the world are experiencing accelerating rates of degradation. These major sources of disturbance impact the biodiversity of fresh waters in many ways. Direct release of chemical toxicants into the surface waters is very common. Many forms of heavy metals, inorganic nutrients and organic compounds enter the aquatic environment. Although some are inactivated by chemical precipitation or oxidation, many have long resident times. Despite dilution and dispersion in the aquatic environment, bio magnification of both metals and organic compounds is common, which increases toxicity exponentially. The biological effects of these compounds are still unknown.

Human impacts on the quality and quantity of fresh water threaten economic prosperity, social stability, and the resilience of ecosystem services and natural capital. As society and ecosystems become increasingly dependent on static or shrinking water supplies, there is increasing risk of severe failures in social or ecological systems, including complete transformation of ecosystems and the possibility of armed conflicts. Rising demand for fresh water can sever ecological connections.

Water quality is dynamic and its changing parameters require the water technologist to be in constant touch with the chemist, the bacteriologist, the biologist and the toxicologist. The importance of the water quality in regard to both human health and production of food supply has become widely recognized in recent years, and to some extent, the importance of water quality in the maintenance of healthy and desirable aquatic population has also joined appreciation.

In recent years, ecologists, hydrologists and engineers have been paying special attention to lotic aquatic environments, as they bear on most important aspects of water use. These now include industrial and agricultural uses impoundments of water for power and flood control, fish and wild life habitat and waste disposal.

The past development and evolution of limnology as a discipline has demonstrated that experimentally controlled disturbances of parts of aquatic ecosystems are essential for quantitative evaluation of casual mechanisms governing their operation. Correlative analysis and modeling only establish hypothesis, not causality, and allow only therapeutic management applications. Rather than constantly searching for

7

differences, commonality must be sought. Among the large diversity of species, communities, and biogeochemical processes controlling growth and reproduction, commonality emerges at the levels of regulation of metabolism.

In an attempt to protect the areas of limnological interest, there is a need to understand the parameters that result in the quantitative physico-chemical and biological status of surface waters.

Another useful biological index that describes water quality conditions is the index of saprobity. Term saprobity includes the pollution of water with organic matters undergoing microbial decomposition. This situation is most common in developed countries and it includes municipal sewage and industrial wastewater. Degrees (levels) of saprobity are defined by communities of organisms (bioindicators) and their range is determined by the saprobic indices. They are regularly present in the saprobiological studies, since they are the only widely accepted means of categorizing water quality.

In India inland water bodies have attracted great attention of various workers leading to the studies on seasonal variations in abiotic and biotic factors.The limnology of tropical lakes is of great interest and has become a main field of contemporary limnological research to overcome the lack of knowledge. Tropical lakes are warm water lakes situated in the tropical and sub tropical parts of Asia, Africa, Central and South America. The chemical and physical properties as well as the biological properties of tropical lakes differ significantly to those of temperate lakes (Lewis 1987), due to thermal stratification and primary production, diversity of fauna and flora in the water bodies. (Gunter Gunkel 2000).

The physical pollution of water brings about changes in water with regard to its colour, odour, density, taste, turbidity and thermal properties etc. The chemical pollution of water causes changes in acidity, alkalinity, P^H, dissolved oxygen and gases in water. The organic pollutants can be biodegradable or non-biodegradable.

Contaminated water supplies frequently create infections of the intestinal tract like dysentery, cholera, typhoid gastroenteritis, and infectious hepatitis.

Any physical, biological and chemical change in water quality that adversely affects living organisms or makes water unsuitable for desired uses is considered to contribute in aquatic pollution. Today water resources have been the most exploited form of natural systems since man strode the earth. Pollution of water bodies is increasing steadily due to rapid population growth, industrial proliferations, urbanization, increasing living standards and wide spheres of human activities.

Many water bodies, rivers of the world receive heavy influx of sewage, industrial effluents, domestic and agricultural wastes which consist of substances varying from simple nutrient to highly toxic hazardous chemicals. Sewage is commonly a cloudy dilute aqueous solution containing mineral and organic matter.

About 75% of water sewage, domestic wastes and food processing plants cause pollution. It also includes human excreta, soap, detergent, metals, glass, rubbish garden waste and sewage sludge form cesspools, etc. industrial effluents discharged into water bodies contain toxic chemicals, hazardous compounds, phenols, aldehydes, ketones, amines, cyanides, metallic wastes, plasticizers, toxic acids, corrosive alkalis, oils, greases, dyes, biocides, suspended solids, non-biodegradable matter, radioactive wastes and thermal pollutants from numerous industries. The principal type of industries, which contribute to water pollution of lakes and rivers in India, are chemicals from pharmaceuticals, coal washeries, soaps and detergents, pulp and paper, sugar, distilleries, tanneries, steel mills, fertilizers, etc.

The main reason of water pollution in India is also the unwise management and over exploitation of the resources. Most of the lakes are made to carry an ever-increasing load of sewage and industrial effluents. Earlier days the lakes, whole water was used to be crystal clear but the discharge of domestic and industrial and agricultural wastes has changed their physical, chemical and biological nature. Fresh water represents a very small part of the total water on earth. The inland water resources, on the surface of earth such as rivers, lakes, reservoirs and ponds became the focus of special attention in the early stage of development of the science of ecology.

9

By definition a natural lake is created due to shallow basin on land in which water from all around accumulates. Reservoirs on the other hand, are man-made impoundments created by damming a lotic system, a stream, or rivers. All over the world such huge reservoirs are constructed mainly to meet irrigational needs. In India after Independence impounding major rivers created massive reservoirs.

Historically, the importance of water management was well understood by rulers in South India. Limnologically speaking a water body is highly influenced by its catchment geology, magnitude of precipitation and anthropogenic influences. No doubt, the system has a capacity to offset alteration in a limited range and maintain its ecological balance. Nevertheless, in the case of impoundments self-regulatory capacity is low. Particularly when a system is enriched due to domestic and industrial wastes the damage is often irreversible. Prior to industrial revolution, aquatic ecosystems were mainly exploited for agriculture and domestic purposes. However, industrialization added a new dimension in the relationship between man and the aquatic ecosystems. In India after 1947, a big boost was given to planned industrialization as it was thought to be a key to progress and prosperity. All over the country big industrial zones were established invariably in the catchments of major rivers and water bodies.

The maintenance of water quality standards in lakes, ponds and reservoirs is necessary so that they do not become nuisance to both aquatic life and man. According to Bourten (1997) without high water quality, nothing of quality will live in it. Prevailing hydrological geomorphic processes govern the physical and prevailing hydrological and geomorphic processes govern chemical features of lakes. The natural chemical loads are derived from precipitation and the dissolved and particulate materials from surface and ground atmospheric gases. Lake water fluctuates in response to surface water runoff, direct precipitation, ground water discharge, rate of evaporation and most importantly human interference. This affects the physicochemical condition of a lake and in turn affects the fauna and flora by imposing physiological and behavioral adaptations. The physicochemical factors are limiting, and their presence or absence produce important and vital consequences on the life of rotifers. A common disturbance to Lake Ecosystem involves progressive increases in loading of nutrients. This increase, particularly among shallow lakes that predominate globally, a marked shift in productivity occurs from the pelagic to surfaces (attached surfaces) associated with living aquatic plants and the particulate detritus of dying (senescing) macrophyte biomass.

Water quality alterations as disturbances to biodiversity: The fresh waters of the world are experiencing accelerating rates of degradation. These major sources of disturbance impact the biodiversity of fresh waters in many ways. Direct release of chemical toxicants into the surface waters is very common. Many forms of heavy metals, inorganic nutrients and organic compounds enter the aquatic environment. Although some are inactivated by chemical precipitation or oxidation, many have long resident times. Despite dilution and dispersion in the aquatic environment, bio magnification of both metals and organic compounds is common, which increases toxicity exponentially. The biological effects of these compounds are still unknown.

Lake restoration efforts have therefore been directed towards reducing the loading of phosphorous and to some extent nitrogen to the surface waters by advanced wastewater treatment, diversion and land management or reducing the phosphorous load in wastewaters by reducing the phosphorous content in detergents. Prevention of nutrient pollution and eutrophication is a long-term solution. Reductions in the nutrient loading require evaluation of diffuse and point sources within the drainage basin and a systematic program of reduction and control.

The structure and composition of macro zooplankton (Cladocerans, Ostracods and Copepods) assemblage are significantly altered with increasing eutrophication, but the response of micro zooplankton (Rotifers and Ciliates) to a trophic gradient is not well documented. Across a variety of lake types protozoa are consistently an abundant component of the planktonic community, and their abundance apparently increases with increasing eutrophy. A few researchers have shown that the species composition and individual abundance of protozoa were strongly related to lake trophic status. Protozoans are consumers of bacterio plankton, phytoplankton and organic matter (Frenchel, 1987; Laybourn-Parry, 1992). They are a very important link in the transfer of energy from bacteria to the higher trophic levels since they are a common diet staple for micro crustaceans and fish larvae (Porter et al., 1985). By reaching very high population densities they play an important role in water layer metabolism, which makes proto zooplankton research essential for understanding the dynamics of the aquatic ecosystems.

Limnology, as a different field of science has developed during the last hundred years. Beginning of knowledge concerning fresh water life, like those of marine life were made in the remote past possibly before the days of Aristotle (384-322 B.C.). However, besides the historical interest involved, no significant scientific contribution waas made for at least nineteen hundred years after the time of Aristotle (Welch 1952).

F.A. Forel (1841-1912), professor in the University of Lausanne, Switzerland recognized the real biological opportunity in the lake investigations and his work became the foundation of modern Limnology. Chumly (1910) listed 116 limnological papers published by Forel during the period (1868-1909) which occupy a significant place in the history of Limnology. Forel work led to the formation of a limnological commission in Switzerland in the year 1890 an international commission and the impact of fresh water investigation began to be felt by establishment of freshwater biological stations in Europe and America. In 1888 Professor Anton Fritsh established the first biological station, in Bohemian Forest, a portable laboratory which used to move to different lakes. Another pioneer station was found by Zacharias (1891) in Germany. Later several freshwater stations were established in Germany, France, Norway, Sweden, Switzerland, Denmark, Austria, Italy, Scotland, Russia, Finland, Belgium and United States. These early freshwater stations yielded a mass of information, which later added materially to the early groundwork of modern ecology.

Before 1870, America lakes were little known biologically and information was very fragmentary though scattered pioneer accounts of the biology of Great Lake appeared from time to time during the first three quarters of the nineteenth century. Agassia (1850) published a volume on "Lake Superior: its physical character, vegetation and animals". Stimson (1870) published a short account on deep-water fauna of Lake Michigan. Smith and Verrill (1871) reported the invertebrates from Lake Superior. In the meantime some interested workers such as Forbe, Brige and Marsh were giving attention to some of the smaller inland lakes. The researchers of Kofoid on the Illinois River, Birge and Juday on Wisconsin lakes, Needham on New York lakes and of many other investigators who worked during this formative period did much to lay a foundation for American limnology. In the recent years extensive and continuous investigations were carried out on fresh water lakes.

Several investigations in Europe contributed to limnological research. Green (1972) on river Sokato and lake Nulanda, Jolly (1965) on New Zealand lakes.

Contributions to hydro biological inland waters have also come from several Asian countries (Bagchi 1986) on Ganga Basin, (Green et.al. 1974) on Indonesian lakes.

Earliest contributions to zooplankton studies in India come from scientists working in the laboratories either of the Zoological Survey of India (Prashad 1916, Sewell 1924, 34 and Pruthi 1933) or with the state or Central Fisheries Departments (Chopra 1930). However, Ganapathi (1959) working as water analyst in Madras City Corporation contributed significantly to ecological studies of hydrobiological significance. However, the bulk of Zooplankton studies in relation to hydrobiology have come only during the last 50 years.Universities have contributed significantly during recent years e.g. Vasisht (1976), Aziz Hussain (1977), Krishna Kumar (1976),I.S. Rao (1982), Jayadevi (1985) Tandon and Singh (1988), Nayar (1970), Mishra et al. (1981), Gopal (1990).

In India studies on several aspects of the fresh water lakes have been carried out by several authors, which include Kaul and Raina (1982), Zutshi (1980), Zutshi and

Khan (1988), Das (2000) Panigrahy (2000) and Gupta et al (2001). Seshavartaram,(1992), Venkata Ramanaiah Solanki (2010)

Smaller water bodies such as tanks and ponds have attracted greater attention because of their easy accessibility and possibility to do intensive studies without costly equipments such as motorized boats, vehicles etc. Thus in the North and North Eastern Parts of India a large number of contributions are available on tanks and ponds such as those of George (1968), Moitra and Bhattacharya (1965), Khan (1969),Shah et.al.(1971),Jana (1973),Tiwari (1977). Similar contributions from South India have come from, Krishnamurthy (1988), Sreenivasan (1969), Michael and Victor (1973), Jyothi and Sehgal (1979).

Lakes and reservoirs have also attracted attention of several workers, Rao et.al. (1976). Hydrobiological studies in hilly areas have also been made in the recent years such as those of Das (1982) on the ecology of Dal lake, Biswas (1964) on the Cladocerans of Simla. Jhingran et.al. (1969) made a useful contribution by together information on methodology based on work in India. Tonapi (1980) has made a very good attempt in presenting the existing state of our knowledge in a comprehensive way on fresh water animals in India.

Due to improper sanitation, industrial waste disposal and faulty water resource management, the percentage of water borne illnesses are increasing day by day. In developing countries, 80 percent infectious diseases are water born and 50 percent of deaths among the children are due to diarrhoeal diseases (Manja and Kaul, 1992). The most common water borne infections are typhoid, cholera, shigellosis, viral associated diarrhea and infectious hepatitis (Pelczar et al., 1988).

The enumeration of total viable count aerobic heterotrophs, total coli forms and E.coli generally indicates the extent of water pollution (Wright, 1986 and Pathak et al., 1988).

Water pollution is due to the alteration in physical, chemical and biological characteristics, which may lead to harmful effect on human and aquatic biota (Abassi et al.1996). It is the addition of excess of undesirable substances to water that makes it harmful for man and aquatic life.

Water pollution is caused due to harmful solids, liquids or gases that are non-permissible, undesirable, unpleasant and objectionable (Shrivastava et al., 2001). Thus water pollution disturbs the normal uses of water for irrigation, agriculture, industries, public water supply and aquatic life.

In contrast to an imposing buildup of literature on temperate waters, not much is known about the tropical reservoirs, particularly those of India. Ever since the pioneering limnological observations of Prasad (1916), several workers have engaged themselves in the hydro biological studies of lakes and ponds found in different parts of India. In general it may be said that such studies have received much attention in southern India, while they remain sadly neglected in other states for a long time. Some of the important contributions, worth mentioning in this regard are those of Kant and Kachroo (1975), Zutshi and Vass (1977),Kal et al. (1978), Vaas (1982) etc.

In India extensive work on different aspects of limnology has been carried out by a number of workers Ganapati, (1943); Zafar (19 68), Munawar (1970), Rao (1975), Chourasia Sadhana (1985) has studied seasonal fluctuations of zooplankton in Burha Tank water, Raipur, M.P. the water quality of lentic ecosystems in Dharwad (Karnataka) is studied by Hegde and Kale (1985). Some of the limnological work in India which is carried out in the recent past and which is worth mentioning are Rao A.M.(2004),Shanty, et al(2002),Eswarlal Sedemekar et al (2003) Usha Awasthi et al (2004) Savita Dixit, et al (2005),) Shiva Kumar et al,(2005),Vijay Kumar(2005) Ramanai Bai (2005),Sanju Sinha and M.L.Naik (2005) Kudari et al (2006) Ravi kumar.M, et al (2006) Padmanabha et al (2006).

Temporary ponds are found throughout the world and they have regional differences in their type and method of formation, many physical, chemical and biological properties. The high variability of abiotic factors and the often comparably variable composition of the fauna, appears particularly important.

The limnology of temporary waters is poorly documented in comparison to that of larger aquatic ecosystems despite their general ecological interest and ubiquitous distribution (Bonner et al. 1997). Williams (1985) pointed out that in spite of their relative ubiquity, little is known about the functional dynamics of these systems and knowledge of the winter period is especially poor.

Numerous workers have examined changes in some physical and chemical factors, but few studies report the seasonal, diurnal and vertical fluctuations of these factors.

Remarkable increases in human population during recent decades have resulted in major changes in the ecology of water bodies. This intensified the use of these water bodies as receiving dumping sites for disposal of an unprecedented amount of human and agricultural effluents. As a result many water bodies have undergone or are undergoing eutrophic or hypertrophic. Such changes are most notable in the water bodies close to populated towns and cities. There are about more than 40 tanks in the Nizamabad District. Andhra Pradesh. Out of those lakes the Bellal Lake is prominent water body situated 2km from the town Bodhan. The Bodhan town is an ancient town. This place was previously known as "Ekachakrapuram", "Bahudanyapuram" and "Podora".

This is a place which ancient associations and inserptions of A.D.1056. The Bodhan town is spread with 21.36 sq. km. The town is located at 18-14 degrees North Latitude and 77-53 degrees East Longitude.

The agricultural practice around the lake Bellal of Bodhan has a direct impact on this water body. The accompanying agricultural practices based on use of fertilizers, pesticides as well as irrigation have been resulting in degradation of water quality both directly by discharging wastes and indirectly by increased demand of water. The domestic and agricultural wastes around Lake margin is a source of pollution in the lake.

Present study embodies results on physico-chemical and biological characteristics of Lake Bellal of Bodhan, a town situated in the North –West part of Andhra Pradesh.

Topo sheets of Bodhan town and Bellal lake (No. 56 F/14 Survey of India) are shown in fig. 1 and fig. 2 respectively.

The lake Bellal has been identified as one of the water Bodies of this region with much biotic potential with respect to planktons. The study has been made to assess the impacts of multiple factors both natural and anthropogenic on monthly variations in physico-chemical complexes as well as biotic components.

The results of this study constitute the first set of data for physicochemical and biological water quality which aid in the development of water quality management of Bellal Lake in Bodhan.

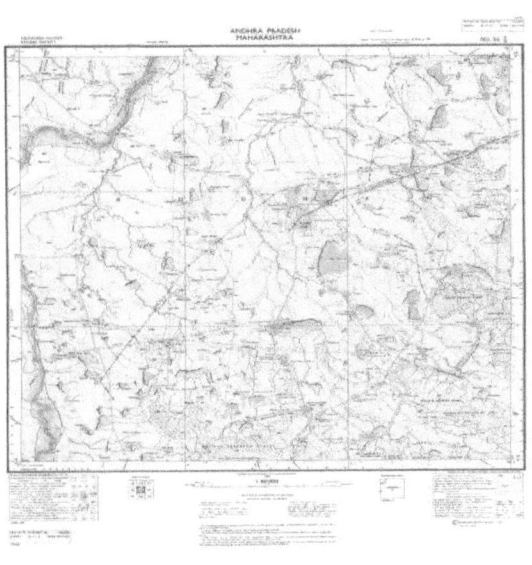

Fig 1 Bodhan Town Map

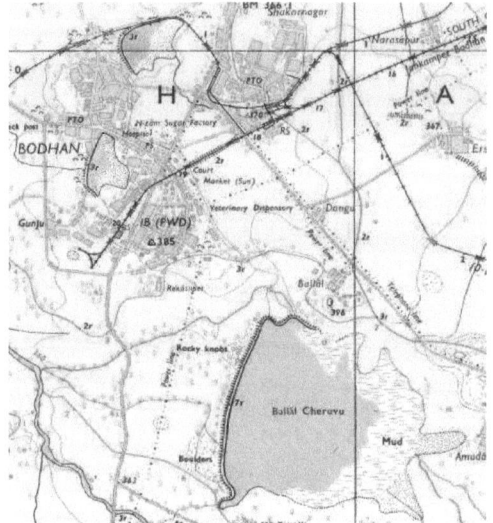

Fig. 2 Bellal lake

Lake morphometry

Bellal Lake

Bellal lake is located on southern side of Bodhan. The total capacity of Bellal Lake is having 346 MCFTcapacity. The main source of water to Bellal Lake is D-40 canal and local streams. The catchment area of Bellal Lake is 46.5 sq miles. Total spreading area of Bellal Lake is 1440 hectares. The depth of Bellal Lake is 12 feet. The water of Bellal is used for domestic (drinking) purpose is 3 cusec/day, agricultural purpose 50 cusec/day and for industrial purpose 10 cusec/day. *Aayakattu* of Bellal Lake is 2000 acres.

Study Area

Fig. 3 Full view of Bellal lake

19

Bellal Station I

Station I is located near the D-40 canal where it receives water from Nizam Sagar. That is inlet to the lake Bellal.

Bellal Station II

Station II is selected for investigation because maximum amount of water is present here and a regular fishing activity is going on at this station.

Bellal Station III

Station III is located because it is out let to the fields, Municipal supply, and also a channel for industrial purposes. These stations are shown in fig. 1.4.

Station I: Bellal Lake at D-40 canal

Station II: Bellal Lake

Station III: Bellal Lake

Fig 4 Different sampling stations of Bellal Lake

Collection of water samples

Water samples for physico-chemical analysis were collected from three stations of Bellal lake. Sites of sampling were identified based on representative stations of the water body so that, by and large the water samples should represent the totality of its water chemistry. Sampling was done once in the second week of each month from August 2002 to July 2004 between 9.00 A.M. to 11 A.M. The samples were taken from 5-8 cm in acid polyethylene bottle of two liter capacity and brought to the laboratory.

The physico-chemical and biological parameters water has been analyzed. Standard methods have been followed for physicochemical analysis as suggested by APHA (1989), Trivedy and Goel (1986).

Material and methods

The present study carried out in Bellal lake of Bodhan, Nizamabad District of Andhra Pradesh. India, during two years starting from August 2002 to July 2004. In the present study various parameters like physico-chemical parameters of water and biological parameters like zooplankton were studied.

The following physico-chemical parameters of Bellal lake was analyzed once in a month to know its chemical and biological status. The parameters analyzed were Temperature, Hydrogen ion concentration (P^H), Total solids (TS), Total suspended solids (TSS), Total dissolved solids (TDS), Total alkalinity, Total hardness, Nitrates (NO_3-N), Phosphates (PO_4-P), Calcium, Magnesium, Chlorides, Dissolved oxygen (DO) and Biological oxygen demand (BOD).

Physico chemical analysis of water samples

Temperature: Monthly water temperature ^0C was taken with the help of thermometer.

P^H: The P^H was determined with the help of digital P^H meter (P^H ep Hanna instrument).

Dissolved oxygen: Winkler's Iodometric Method

Principle: The manganous sulphate reacts with the alkali (KOH or NaOH) to form a white precipitate of manganous hydroxide which in the presence of oxygen gets oxidized to a brown colour compound. In the strong acid medium manganese ions are reduced by iodide ions which get converted into iodine equivalent to the original concentration of oxygen in the sample. The iodine can be titrated against thiosulphate using starch as an indicator.

Procedure: Sample was filled in a glass stoppered bottle of known volume (100-300 ml) carefully, avoiding any kind of the air bubbles in the bottle after placing the stopper. 1 ml of $MnSO_4$ and alkaline KI solutions were poured well below the surface from the walls. The reagents can also be poured at the bottom of the bottle with the help of special pipette syringes to ensure better mixing of the reagents with the sample. A precipitate was observed. Stopper was placed and shaken the contents well by inverting the bottle repeatedly. The bottle was kept some time to settle down the precipitate. 1-2 ml of concentrated H_2SO_4 added and shaken well to dissolve the precipitate.

The whole contents were removed in a conical flask for titration. within one hour of dissolution of the precipitate contents were titrated against sodium thiosulphate solution using starch as an indicator. At the end point, initial dark blue colour changes to colorless.

Calculations

Dissolved oxygen, ppm = Normality of titrant sodium thiosulphate solution $\times 8 \times 1000 /$ Volume of the part of the contents titrated – volume of $MnSO_4$ and KI added

Biological Oxygen Demand

Principle: Biological oxygen demand (BOD) is the measure of degradable organic material present in a water sample, and can be defined as the amount of oxygen required by the micro organisms in stabilizing the biologically degradable organic matter under aerobic conditions. The principle in the method involves, measuring the difference of the oxygen concentration between the sample after incubating it for five days at 20 ^0C.

Procedure: Water dilution was prepared in a glass container by bubbling compressed air in distilled water for about 30 minutes. 1 ml each of phosphate buffer, magnesium sulphate, calcium chloride, and ferric chloride solutions were added for each liter of dilution water and mix thoroughly. The sample was neutralized to P^H 7.0 by using IN NaOH or H_2SO_4. Since the dissolved oxygen in the sample likely to be exhausted; it is usually necessary to prepare a suitable dilution of the sample according to the expected BOD range.

Dilutions were Prepared in bucket or a large glass trough and were mixing the contents thoroughly. Two sets of the BOD bottles were filled. one set of the bottles in BOD incubator at 20 ^0C for 5 days was kept, and determined the dissolved oxygen content in another set immediately.

The dissolved oxygen was determined in the sample bottles, immediately after the completion of 5 days incubation.Similarly for blank, 2 BOD bottles were taken for dilution water. In one, the dissolved oxygen content was determined and the other incubated with the sample to determine the dissolved oxygen after 5 days.

Calculation

BOD, Mg/L= initial dissolved oxygen in the sample- dissolved oxygen after 5 days× dilution factor

Total alkalinity

Principle: Alkalinity is determined by titrating the water sample with standard solution of strong acid. The equivalency or end points of the titration are selected as the inflection points in the titration of sodium carbonate with H_2SO_4. These points will vary slightly with temperature, ionic concentration, and free carbon dioxide concentration. However, the points are usually taken as P^H 8.3 for the carbonate end point P^H 4.5 bicarbonate end point.

Procedure: 100 ml of sample is taken to which few drops of phenolphthalein indicator was added. If the solution remains colour less, phenolphthalein alkalinity was assessed and carbonates were nil. To the same water sample two drops of methyl orange were added and sample was titrated against acid till the yellow colour turned pink. Phenolphthalein alkalinity and total alkalinity were calculated as per formulae given below

Phenolphthalein alkalinity in ppm =Titer value ×Normality of titrant × 1000×50/ ml of sample
(Titer value for the volume of acid to decolourize phenolphthalein)
Total alkalinity in ppm = Titer value ×Normality of titrant ×1000×50/ ml of sample
(Total amount of acid used for converting the sample into pink colour)

Total hardness

Principle: Calcium and magnesium form a complex of wine red colour with Erichrome Black T at P^H of 10.0±0.1. The EDTA has got a stronger affinity towards Ca^{++} and Mg^{++} and therefore by addition of EDTA, the former complex is broken down and a new complex of blue colour is formed.

Procedure: Hardness was examined by EDTA titrometric method. To 50 ml of water sample, 1 ml of Buffer solution and 100 mg of Erichrome black T indicator were added and then titrated against 0.01 M EDTA solution till the wine red colour changes to blue. Hardness was calculated as follows

Calculation

Hardness in ppm = Titer value of EDTA×0.01×1000/ ml of sample

Calcium

Principle: Many indicators such as Muroxide, calcon, etc. form a complex with only calcium but not with magnesium at higher PH. As EDTA is having a higher affinity towards calcium, the former complex is broken down and magnesium is precipitated out as magnesium hydroxide. Now calcium reacts with EDTA to form the coloured complex.

Procedure: To 50 ml of sample, 2 ml of 1N NaOH solution and 100 mg of Muroxide indicator were added.

It was titrated with 0.01 M EDTA solution till pink colour changes to dark purple. Calcium was calculated as follows

Calculation

Calcium in ppm = ml of EDTA used×0.01×1×400.8/ ml of sample

Magnesium: colorimetric method (brilliant yellow)

Principle: when magnesium hydroxide is precipitated in the presence of brilliant yellow, the dye is adsorbed on the precipitate and its colour changes from orange to red. A stabilizer is added to maintain the Mg(OH)₂ in colloidal suspension.

25

Procedure: In the 100 ml measured volumetric flack 50 ml of sample is taken and 1 ml sodium sulfite solution is added. In order 1 ml of sulphuric acid solution, 20 ml of calcium sulphate solution and 5 ml of aluminium sulphate reagent were added. Volume was brought to about 80 ml with distilled water. 5 ml stabilizer, 2.0 ml brilliant yellow solution and 3.5 ml NaOH were added and diluted to the mark, shaken well and was left for five minutes for full colour development. Sufficiently diluted solution was measured photometrically within one hour, against a blank prepared from distilled water and all of the reagents. Use a light path of 2 cm or longer. Prepare standard curve, from 0, 100, 200, 400 and 600μg/Mg. Check the least two points with each set of determinations.

Calculation

$$mg/L \ Mg = \frac{\mu g \ Mg}{ml \ sample}$$

Chloride

Principle: Silver nitrate reacts with chloride to form very slightly soluble white precipitate of AgCl. At the end point when all the chlorides are exhausted free silver ions react with chromate to form silver chromate of reddish brown colour.

Procedure: Chloride in water samples was estimated by titrating 50 ml sample against 0.02 N $AgNO_3$ solutions using 2 ml of 5% Potassium chromate (K_2CrO_4) as an indicator. At the end point yellow colour changes to persistent reddish brown.

calculation

Chloride (ppm) = Titer value ×Normality of titrant ($AgNO_3$)×1000×35.5/ ml of sample

Phosphates

Principle: The phosphates in water react with ammonium molybdate and form complex heteropoly acid (molybdophophoric acid) , which gets reduced to a blue colour complex in the presence of $SnCl_2$. the absorption of light by this blue colour can be measured at 690 nm to calculate the concentration

Procedure: To 50 ml of sample, 2 ml of ammonium molybdate solution and 5 drops of Stannous Chloride were added due to which blue colour developed. After 5 minutes and before 12 minutes its optical density was

taken at 690nm. The concentration of phosphate was determined with the help of standard curve.

Nitrates

Principle: Nitrate and brucine react to produce a yellow colour, the intensity of which can be measured at 410 nm. The reaction is highly dependent upon the heat generated during the test. However, it can be controlled by carrying out the reaction for a fixed temperature. The method is suitable for the samples having very wide range of salinity.

Procedure: 50 ml of sample was taken and was placed on cool water bath and 2 ml of NaCl and 10 ml of H_2SO_4 were added after mixing thoroughly 0.5 ml of Brucine reagent was added and placed on hot water bath for 20 minutes. The contents were cooled again in a cold water bath and readings were taken at 410 nm. Prepared standard curve between concentration and absorbance by taking the dilutions from 0.1 to 1.0 mg N/L at the interval of 0.1.

Total Solids (T.S)

Principle: Total solids are determined as the residue left after evaporation of the unfiltered sample.

Procedure: An evaporating dish of about 100 ml capacity was taken and ignited at 550 ± 50 ^0C in a muffle furnace for about an hour, there after it was cooled in a desicator and weighed.

100 ml of unfiltered sample was evaporated in the evaporating dish on a hot plate having temperature not more than 98 ^0C. The residue was heated at 103-105 ^0C in an oven for one hour and taken the final weight after cooling in a desicator.

Calculation

Total solids (ppm) = Final weight of dish in grams- initial weight of dish in grams/ Volume of sample evaporated in ml

Total dissolved solids (TDS)

Principle: Total dissolved solids are determined as the residue left after evaporation of the filtered sample.

Procedure: An evaporating dish of about 100 ml capacity was taken and ignited at $550\pm50\ ^0C$ in a muffle furnace for about an hour, there after it was cooled in a desicator and weighed. The sample was filtered through glass filter paper by applying the suction.

100 ml of unfiltered sample was evaporated in the evaporating dish on a hot plate having temperature not more than $98\ ^0C$. The residue was heated at $103\text{-}105\ ^0C$ in an oven for one hour and was taken for the final weight after cooling in a desicator.

Calculation

Total dissolved solids (ppm) = Final weight of dish in grams- initial weight of dish in grams/ Volume of sample evaporated in ml

Total suspended solids (TSS)

Total suspended solids were determined as the difference between total solids and total dissolved solids.

T.S.S.was calculated as T.S.S. (ppm) = Total solids – Total dissolved solids

Zooplankton

Biological examination of water was done by collecting 20-25 liters of water from each site. The water was collected by sampler and filtered through net . In total around 100 liters of water were collected from each lake, each month, from which the zooplankton studies were made. The zooplanktons copepods, cladocerans, rotifers, ostracods and certain larvae were drained out from the planktonic sampler which had been accumulated in the sample cylinder. They were fixed 4 % formalin and identified to the genus level with the help of fresh water biology Edmondson(1965). Counting of organisms was done using Sedgwick- Rafter counter and the dilution technique and the population density of each organisms is represented per liter of water.

The results obtained were analysed statistically.

Results and Discussions

Temperature

Temperature is primary environmental factor that affects the activities of biological species and solubility of gases in water. It is noticed that water temperature is always lower than that of air temperature due to various reasons like gases in the air, humidity, dust and other colloidal particles. In early hours i.e., before photosynthetic activity starts, temperature is slightly more on surface water than in deep water.

In the present work, water temperature of three sampling stations of Bellal lake, during study period ranged between 20 °C to 35.8 °C. It was observed that the maximum temperature was in the month of May and minimum was in the month of January. Monthly variations of temperature and mean values were shown in the Fig. 5 and Table.1 respectively.

Temperature is one of the most important ecological factors, which controls the physiological behaviour and distribution of organisms. High temperature was well marked in summer, while low in winter. This is due to seasonal atmospheric variation in temperature. Welch (1952) has observed that shallow water reacts more quickly to atmospheric changes. The temperature ranged between 23°C to 35°C. There is no significant variation in temperature between the different stations of Bellal Lake.

Station	N	Mean	Std. deviation	Std.error
1	24	26.7542	3.5285	.7203
2	24	26.5292	3.7832	.7722
3	24	26.7417	3.7339	.7622
Total	72	26.6750	3.6327	.4281

Table 1 Mean values of Temperature °C
of Bellal Lake

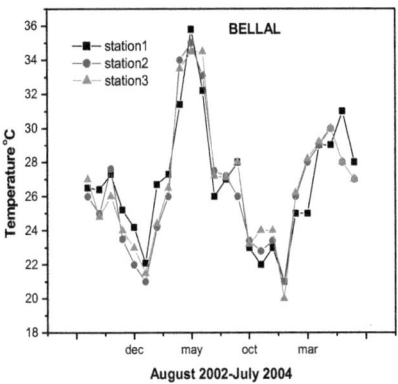

Fig 5 Monthly variations in

Temperature °C

in Bellal Lake

P^H

P^H is indicative of hydrogen ion concentration and it expresses the intensity of acidity and alkalinity of water. Any change in P^H of water is accompanied by changes in the other physico- chemical aspects of the medium.

In the present investigation the P^H of Bellal Lake at all the stations ranged between 7.2 and 8.9. indicating

alkaline nature of the Bellal Lake water. The P^H ranged between 7.2 and 8.9 with an average of 8.27 ± 0.42 at station I, while at station II it was7.40 to 8.60 with an average of 8.19±0.31 and 7.6 to 8.80 with an average of 8.28 ± 0.36 at station III. The maximum and minimum values were recorded during different months. The average values are depicted in Table. 3.2 and monthly variations are given in the Fig. 3.2. In the present study, P^H of water was alkaline throughout the study period. This is in accordance with earlier reports by Neera Srivastav et al, (2003), who reported that the value of P^H ranges from 7.2 to 9.0. There are no significant variations in P^H at different stations of Bellal Lake. The P^H values of Bellal Lake can be explained as moderately alkaline in its nature. Moderate P^H in Lake Bellal is due to the dilution

31

effect during rainy months. These higher values may be due to the increased photosynthetic activity because maximum phytoplankton density contributes to flow of organic materials into the lake (Hussain et al., 2004).

Station	N	Mean	Std. Deviation	Std. Error
1	24	8.2792	.4293	8.764 E-02
2	24	8.1958	.3141	6.412 E-02
3	24	8.2833	.3608	7.364 E-02
Total	72	8.2528	.3681	4.338 E-02

Table 2 Means values of P^H in Bellal Lake

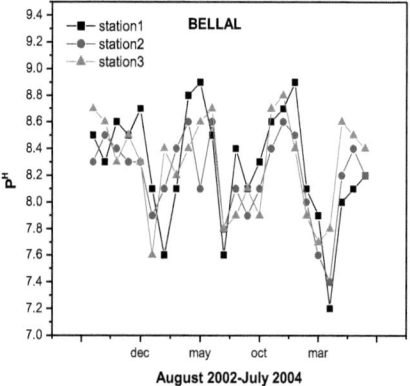

Fig 6 Monthly variations of P^H in Bellal Lake

Dissolved oxygen

Dissolved oxygen is one of the most important parameters in assessment of water quality and reflects the physical and biological process prevailing in the waters. In the present study dissolved oxygen ranges from 4.40 to 13.70 ppm in Bellal Lake with an average 8.41±3.05ppm.

Dissolved oxygen was observed to be lowest in concentration in August 4.4ppm at station I and it has been found to be 5.2 ppm at station II, 4.8ppm at station III. It was maximum in the month of February with 13.1ppm at Station I, 13.6ppm at station II and 13.7ppm at station III in the month of January. Monthly variations are shown in the Fig. 7 and mean values are given in the Table.3.

Super saturation of dissolved oxygen in Lake Bellal was observed most probably as a result of photosynthetic activity in the surface water of the lake. The dissolved oxygen observed to be more at station II as compared with Station I and Station III. Dissolved oxygen was found to be maximum in the month of January.

When the water temperature was low the values of dissolved oxygen were fairly high. Thus it has an inverse relationship with dissolved oxygen. However, Sreenivasan (1974) reported that the water temperature did not affect the dissolved oxygen. Relatively higher values of dissolved oxygen might be due to increased solubility of oxygen at low temperature (Chatterjee and Raziuddin, 2002). In the present study the dissolved oxygen showed a negative correlation with temperature. Masood Ahmed and Krishna Murthy (1990) showed a positive correlation between temperature and soluble gases like dissolved oxygen. In the present study no significant variation in dissolved oxygen was observed between the stations I with II and III, II with I and III, and III with II and I of Bellal Lake.

Station	N	Mean	Std. Deviation	Std. Error
1	24	8.0292	3.1363	.6402
2	24	8.6125	3.0306	.6186
3	24	8.6042	3.0797	.6286
Total	72	8.4153	3.0512	.3596

Table 3 Mean values of Dissolved

Oxygen ppm in Bellal Lake

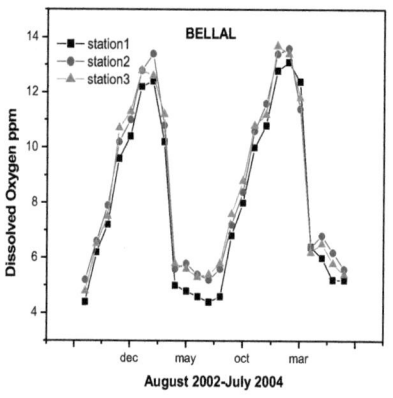

Fig7: Monthly variations of

Dissolved Oxygen ppm in

Bellal Lake

Biological Oxygen Demand (BOD)

Biological oxygen demand is the amount of oxygen utilized by microorganisms in stabilizing the organic matter. On an average basis, the demand for oxygen is proportional to the amount of organic waste to be degraded aerobically. In the present work, during the period of investigation BOD varied from 13.80 to 40.20 ppm with an average 25.39± 7.71 ppm in Bellal Lake. The maximum values were recorded in

34

June 2003 at all the stations and minimum values were in July 2004. The average values are given in the Table 3.4 and monthly variations are depicted in Figure 3.4. From this figure it is clear that the BOD observed is less at station II as compared with the other two stations. The observed increase in Biological oxygen demand value might be due to biological activities at elevated temperatures. BOD is the major criteria used in stream pollution control where organic loading must be restricted to maintained desired DO. In unpolluted waters BOD is lower while it is high in the case of polluted waters (Hussain et al, 2004).

Monthly variation of BOD which ranged between 13.80 ppm to 40.20 ppm indicates the slow eutrophication

phase of the lake as supported by Devaraju et al (2005) in Maddur Lake. It is negatively correlated with P^H and the values are significantly different. The significant variations were observed between the station II with I and III of Bellal Lake

Station	N	Mean	Std. Deviation	Std. Error
1	24	25.7667	7.8095	1.5941
2	24	24.9250	7.7673	1.5855
3	24	25.4833	7.8771	1.6079
Total	72	25.3917	7.7152	.9092

Table 4: Mean values of Biological oxygen demand ppm in Bellal Lake

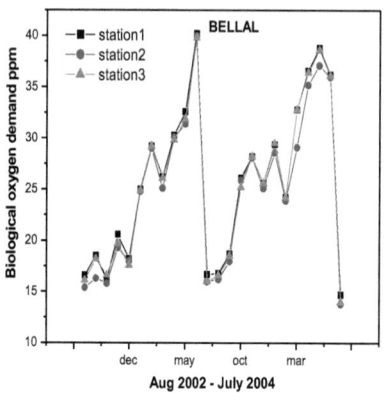

Fig8: Monthly variations of

Biological oxygen demand

ppm in Bellal Lake

Total Alkalinity

The alkalinity in water is mainly due to carbonates and bicarbonates and partially due to hydroxides and silicates. Alkalinity in itself is not harmful to human beings; still the water supplies with less than 100 ppm are desirable for domestic use. Expected total alkalinities in nature usually range from 20 to 200 ppm.

Total alkalinity in Bellal Lake ranged from 116 to 184 ppm, monthly average values of recorded as 153.08± 3.09 ppm at station I, 163±2.97 ppm at station II, and 144.08±3.01 ppm at station III. Average mean values of alkalinity of different stations of Bellal Lake are presented in Table. 5 and monthly variations are depicted in Figure9.

The lowest value noted was during August 2002 and maximum value recorded was during July 2003. This observation is similar to that of Usha Awasthi and Sunil Tiwari (2004). It was observed that the total alkalinity was highest during summer months of two years study.

Total alkalinity declines during rainy months may be because of rainfall, which leads to the dilution of ionic content of the water body (Shastri and Pendse 2001). In the

36

present study, the values of total alkalinity are significantly different at 0.05 levels between stations I with II and III, and station III is difference with the both stations station I and II. The total alkalinity of Bellal Lake showed a negative correlation with P^H of Bellal Lake. This is supported by Prithwiraj Jha and Sudip Bharat (2003).

Station	N	Mean	Std. Deviation	Std. Error
1	24	153.0833	15.1483	3.0921
2	24	163.0000	14.5632	2.9727
3	24	144.0833	14.7498	3.0108
Total	72	153.3889	16.5543	1.9509

Table 5: Mean values of Total Alkalinity ppm in

Bellal Lake

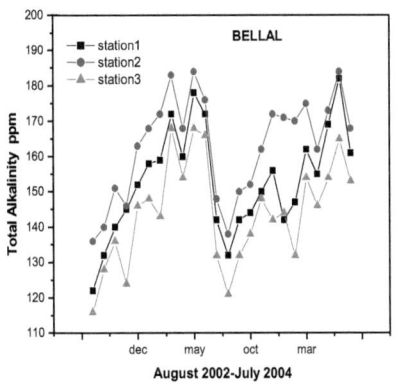

Fig 9.Monthly variations of

Total Alkalinity ppm

in Bellal Lake

Total hardness

Hardness is the property of water which prevents the lather formation with soap and increases the boiling point of water. Principle cations imparting hardness are calcium and magnesium. Highest desirable level of hardness is 100 ppm and maximum permissible level of hardness is 500 ppm (ISI).

Hardness has no adverse effects on health; however, some evidence has been given to indicate its role in heart disease (Peter1974). The range of total hardness in the present lake was 70.0 ppm to 196.00ppm average 124.19±29.15 ppm. Variations in the monthly average values of total hardness in the lake was 70.00 ppm to 196 ppm with an average value of 128.66±32.46 at Station I, 84 ppm to 184 ppm with an average value of 128.00±27.85 at Station II and 72.00 ppm to 182 ppm with an average value of 115.91±26.21 at Station III. Its lowest values were recorded during April and highest values were in the month of March. Average mean values of hardness of different stations of Bellal Lake are presented in Table.6. and monthly variations are depicted in Fig. 10.

Station	N	Mean	Std. Deviation	Std. Error
1	24	128.6667	32.4689	6.6277
2	24	128	27.8521	5.6853
3	24	115.9167	26.2147	5.3511
Total	72	124.1944	29.1589	3.4364

Table 6 Mean values of

Total hardness

ppm in Bellal Lake

The hardness values of water of Bellal Lake were found to be fluctuating positively with respect to total alkalinity. The higher values of total hardness of water is may be due to evaporation of water and addition of calcium and magnesium salts. This is also supported by Bagde and Varma (1985). Hardness below 300 ppm is considered potable but beyond this limit produces gastro intestinal irritation. (ICMR 1975). The statistical mean difference was observed in the present investigation between station II with I and station III with I and II.

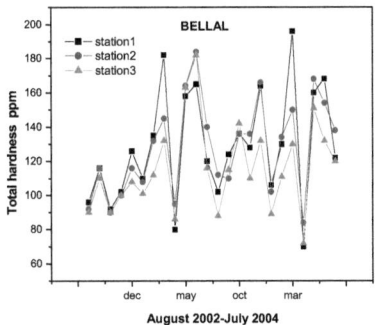

Fig10: Monthly variations of

Total hardness ppm in

Bellal Lake

Calcium

Calcium is one of the most abundant elements of the natural waters. Being present in high quantities in the rocks, it is leached from there to contaminate the water. The quantities in natural waters generally vary from 10 to 100 ppm depending upon the types of rocks.

Calcium as such has no hazardous effects on human health. In fact, it is one of the important nutrients required by the organisms. Concentrations up to 1800 ppm have been found to be not impairing any physiological activity in human beings. (Leher et al 1980).

Calcium content of Bellal Lake fluctuated between 21.66 ppm to 66.44ppm with an average of 39.86 ppm in Bellal Lake. The values of calcium were 24.39ppm to 66.44 (average 40.21±13.21) at station I, 23.54 to 64.76 ppm (average 39.52±12.16) at station II and 21.86to 65.60-ppmit (average 39.87±12.41) at station III. The minimum values were obtained in August 2002 and August 2003 at station I and II and the minimum values were recorded at station III in the month of September of 1st year and August of 2nd year. The maximum values were obtained in the month of May of both years at all sites. Monthly variations of calcium are shown in Figur.11 and average values are given in the Table 7.

Water with calcium values above 25ppm is classified as calcium rich. In the present study, the values of calcium content were above this limit for a major period of the year. Slight rise in calcium content in summer at all the stations could be attributed to the rapid oxidation of organic matter, while it decreased during rainy months, might be on account of dilution due to the rain water into the water body. The values of calcium content show different statistics between the station II with I and III.

and the calcium values were negatively correlated with total hardness. Awasthi and Tiwari (2004). Calcium content in the lake water varies habitat to habitat.

Station	N	Mean	Std. Deviation	Std. Error
1	24	40.2042	13.2125	2.6970
2	24	39.5292	12.1667	2.4835
3	24	39.8754	12.3627	2.5235
Total	72	39.8696	12.4133	1.4629

Table 7 Mean values of

Calcium ppm in Bellal Lake

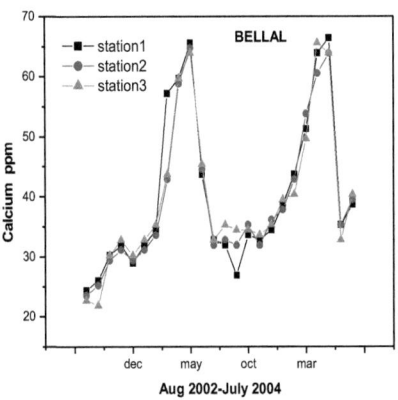

Fig 11: Monthly variation in
Calcium ppm in Bellal Lake

Magnesium

Magnesium also occurs in all kinds of waters along with calcium, but its concentration remains generally lower than the calcium. Magnesium is supposed to be nontoxic at the concentrations generally met with in natural waters.

Desirable limit of magnesium in natural water is 30 ppm and permissible level is 150 ppm. Concentrations as high as 500 ppm impart an unpleasant taste to the water thus rendering it unpalatable.

In the present study the magnesium content varied between 32.74 to 99.46ppm in Bellal Lake. It was recorded to be maximum in May 2003 with 98.2 ppm at station I, 96.94 ppm at station II and 95.69 at station III, while the lowest values were recorded in August 2002 with 36.51 ppm, 34.41 ppm and 33.99 ppm at station I, II and III respectively. At station I magnesium content of water showed successive increasing from August to May, while during rainy seasons it decreased at all the sites.The average values are given in Table.8 and monthly variations are shown in the Fig. 12.

During the present study magnesium concentration was found to be higher than the calcium. According to Awasthi and Tiwari (2004) the magnesium values were recorded always lower than calcium.

Magnesium adds to the hardness of the water along with calcium and causes the problem of scale formation in the boilers. Like other elements magnesium was also found dissolved in water and influenced fauna. In the present study the magnesium is negatively correlated with P^H and dissolved oxygen.

Station	N	Mean	Std. Deviation	Std. Error
1	24	60.1763	19.7941	4.0405
2	24	59.1375	18.2674	3.7288
3	24	60.1483	18.3691	3.7496
Total	72	59.8207	18.5625	2.1876

Table 8: Mean values of Magnesium
ppm in Bellal Lake

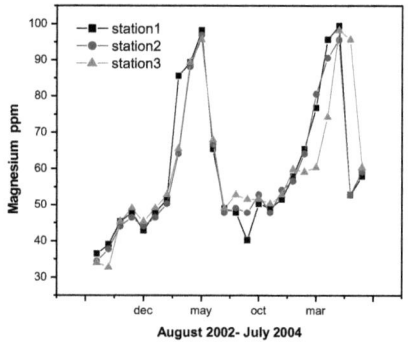

Fig 12: Monthly variations of Magnesium
ppm in Bellal Lake

Chlorides

Chlorides occur naturally in all types of waters. In natural fresh waters, however, its concentrations remain quite low and are generally less than that of sulphates and bicarbonates. It is harmless up to 1500 ppm concentrations but produces salty taste at 200-250 ppm level.

In the present investigation, the chloride content varied from 63.82 to 226.92 ppm at station I, 70.91 to 241.11 ppm at station II and 85.10 to 205.65 ppm at station III. Maximum concentration of chlorides was recorded in May 04 at station I and station II and in June at station III, while minimum was recorded in August at all the stations. Average mean values are given in Table. 9 and monthly variations are shown in Fig. 13

The chloride concentration depends on the characteristic of the sediment and pollution load. The chloride content gradually increased from winter and reached maximum during the month of June and minimum values were observed in the August. Salodia (1996) has reported the similar results, while studying the chlorides in Jaith sagar lake, Rajasthan. However, present findings do not support the observations made by Singh (1979). Low concentrations of chlorides may be attributed to dilution by rapid in flow of water (Chattarjee 2002).

According to Awasthi and Tiwari 2004, the chloride content was more in the samples studied by them. The higher concentration of chlorides is considered to be an indicator of pollution due to higher organic waste of animal origin (Shanti et al, 2002).

Station	N	Mean	Std. Deviation	Std. Error
1	24	158.2174	40.3952	8.2456
2	24	166.1235	34.9471	7.1336
3	24	164.8817	33.4988	6.8379
Total	72	163.0742	36.0552	4.2491

Table 9: Mean values of Chlorides ppm in

Bellal Lake

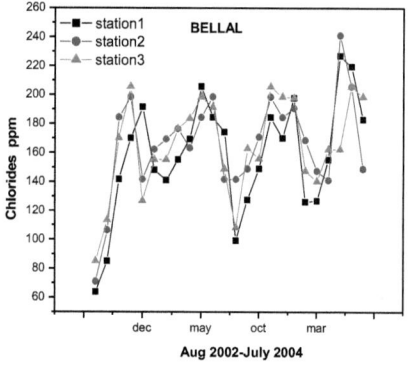

Fig 13: Monthly variations of Chlorides

ppm in Bellal Lake

Phosphates

Phosphorus in the natural fresh waters is present mostly in inorganic forms such as $H_3PO_4^-$, HPO_4^{-2}, and PO_4. Phosphorus an important constituent of biological systems, may also be present in the organic forms. In natural water phosphates are present in small quantities.

Generally aquatic ecosystems receive excess of this nutrient through untreated domestic sewage, agricultural effluents with fertilizers and industrial wastes. The higher concentration of phosphates, therefore, is indicative of pollution.

Phosphates varied from 0.03 to 0.10 ppm with an average of 0.065 ± 0.017 ppm at Station I; 0.04 to 0.09 ppm with an average 0.05 ± 0.015 and 0.03 to 0.09 ppm with an average 0.05 ± 0.016 at stations II and III respectively. It has been found to be maximum in June and minimum in October at Station I, maximum in March 2004 with a quantity of 0.09ppm and minimum in the month of August at Station II and maximum 0.09ppm and minimum with 0.036 at Station III in August.

Throughout the investigation period the PO_4-P concentration was found to be low. The maximum value of 0.10 ppm was obtained at station I in June 2004. A minimum value of 0.03 ppm was recorded at station II and station III.

The same levels of phosphate concentration have been reported by Prithwiraj Jha (2003). Phosphate is considered amongst the primary limiting nutrients in ponds and lakes. The present values differed from the findings reported by Savitha Dixit et al (2005). Higher values of this nutrient at station I may be due to discharge of agricultural runoff compared to other sites. The values are significantly different with 0.05ppm levels between I with II and III, II with I and II, and III with I & II. The same values were reported by Ravikumar et al (2006). According to Ravi Kumar et al (2006) there is increase of PO_4 concentration during winter months in Kullahalitank Karnataka. In the present investigation, the PO_4 concentrations increased during rainy season and are negatively correlated with P^H and total hardness.

The monthly variations of phosphates in the present study are shown in the Fig 14.and average values are given in the Table. 10.

45

Station	N	Mean	Std. Deviation	Std. Error
1	24	6.575 E-02	1.739 E-02	3.550 E-03
2	24	5.354 E-02	1.559 E-02	3.183 E-03
3	24	5.463 E-02	1.677 E-02	3.423 E-03
Total	72	5.797 E-02	1.728 E-02	2.037 E-03

Table 10 Average values of

Phosphates ppm in Bellal Lake

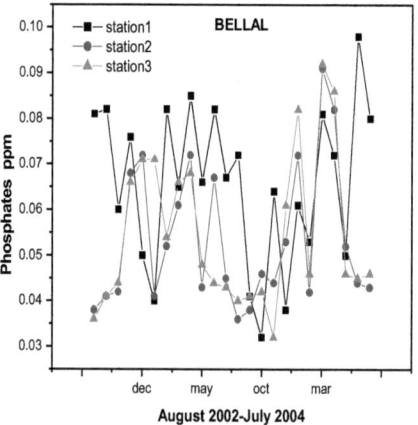

Fig 14: Monthly variations of Phosphates

ppm in Bellal Lake

Nitrates NO$_3$-N

Monthly nitrates varied from 0.62 to 1.60 ppm with an average value of 1.04±0.28 ppm for station I, 0.69 to 1.78 ppm with an average value of 1.10±0.29 ppm at station II and 0.74 to 1.89ppm with an average value of 1.11± 0.30 ppm at station III. Maximum values were recorded in June 2004 i.e., 1.60ppm, 1.78ppm and 1.89ppm at station I., II and III respectively and lowest values are in the month of August 2003 at all the stations. The monthly variations of nitrates of Bellal Lake at different stations are shown in the Fig. 15.

The main source of the nitrate is the nitrification of ammonia and nitrites in the nitrogen cycle and agricultural runoff and decomposition of organic fertilizers.

The present levels of nitrates were similar to that in Ooty Lake (Thilaga et al 2005). The NO$_3$-N is negatively correlated with P^H and dissolved oxygen (Kudari et al 2006).

Nitrates in water when accumulated in soil lead to problem of nitrate pollution in crops like leafy vegetables when the plants grow in soils rich in nitrate develop tumors (Majumdar 2000). High levels of nitrate content stimulating algal growth to such an extent that the water may not be suitable for human consumption. Prithwi (2003) also had similar point of view while working in Mirik Lake. There is no significant difference between the stations of Bellal Lake. The average mean values of nitrates in the present investigation are given in the Table.11.

Station	N	Mean	Std. Deviation	Std. Error
1	24	1.5921	.1608	3.283E-02
2	24	1.6004	.1094	2.232E-02
3	24	1.5700	.1329	2.713E-02
Total	72	1.5875	.1347	1.587E-02

Table 11: Average values of Nitrates

ppm in Bellal Lake

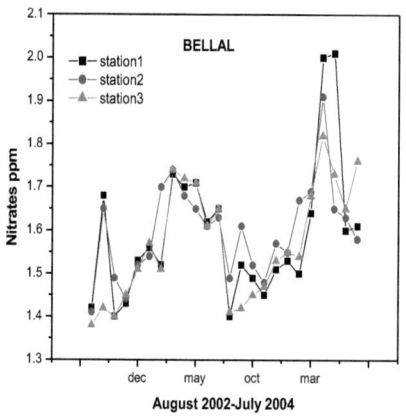

Fig 15: Monthly variation of Nitrates

ppm in Bellal Lake

Total solids

Total solids in the lake ranged from 303 to 437 ppm for station I, 342 to 443 ppm for station II and 354 to 446 ppm for station III. The mean values of total solids were 372.20 ± 7.69, 387.41 ± 6.43 and 403.58 ± 5.74 ppm for station I, station II and station III respectively.

48

The maximum values of total solids recorded were in June 2004 i.e. 437 ppm at station I, while minimum was in August 2002 at 303 ppm. At station II maximum was 443 ppm in July 2003 and a minimum of 342 ppm in August, 2002 and March 2003 at station III the maximum total solids was in the month of December 2003 at. 446 ppm and minimum 354 ppm in August 2002. The monthly variations of total solids in the present study are shown in Fig.16 and the average values are given in the Table. 12.

The findings are not coinciding with the observations made by Pawar et al (2006). The total solids are inversely proportional to temperature and directly proportional to levels of calcium and magnesium,

phosphates and nitrates. This is as observed by Gongalves and Joshi (1946). There is a significant difference of total solids between the different stations studied at 0.05 levels.

Station	N	Mean	Std. Deviation	Std. Error
1	24	372.2083	37.7059	7.6967
2	24	387.4167	31.5070	6.4313
3	24	403.5833	28.4130	5.7998
Total	72	387.7361	34.7863	4.0996

Table 12: Average values of Total

Solids ppm in Bellal Lake

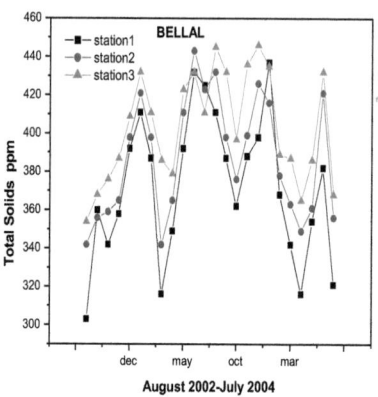

Fig 16: Monthly variations of Total

Solids ppm in Bellal Lake

Total Dissolved Solids (TDS)

Dissolved solids denote mainly the various kinds of minerals present in the water. However, if some organic substances are also present in polluted waters, they may also contribute to the dissolved solids. In Bellal lake total dissolved solids ranged from 280 to 431ppm. The total dissolved solids are 280 ppm to 421 ppm with an average 348.62 ± 37.69 at Station I, 322 ppm to 423 ppm with an average of 366.79 ± 30.89 at Station – II and 337 ppm to 431 ppm with an average of 384.83 ± 31.20. The maximum values were recorded in the month July, 2003 at all the stations, it was minimum in August 2002 at all the stations. Total dissolved solids variations of Bellal Lake are shown in Fig 17.

Average values are given in Table 13.Total dissolved solids are negatively correlated with water temperature, calcium, magnesium, phosphates and nitrates. Significant difference is observed between different stations of the Bellal Lake.

Station	N	Mean	Std. Deviation	Std. Error
1	24	348.6250	37.6919	7.6938
2	24	366.7917	30.8967	6.3068
3	24	384.8333	31.2029	6.3693
Total	72	366.7500	36.1448	4.2597

13: Average values of

total dissolved

solids ppm in Bellal Lake

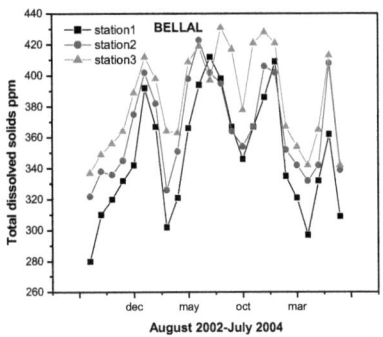

Fig 17: Monthly variations of

Total dissolved solids ppm in

Bellal Lake

Total Suspended Solids (TSS)

In the present investigation the total suspended solids varied between 12 to 50 ppm in the lake water.   At Station I, the range was 12 to 50 ppm with an average of 23.58 ± 10.39ppm, 13 to 37 ppm with an average 20.62 ± 6.27ppm at Station II and at Station III 13 to 33 ppm with an average of 18.75 ± 4.74ppm. The minimum values recorded in July 2004 at all the stations while the maximum values in December 2002 in Bellal Lake. Total suspended solids are negatively correlated with temperature, calcium, magnesium, phosphates and nitrates. Significant difference is observed between the different stations.

Station	N	Mean	Std. Deviation	Std. Error
1	24	23.5833	10.3920	2.1212
2	24	20.6250	6.2750	1.2809
3	24	18.7500	4.7480	.9692
Total	72	20.9861	7.6848	.9057

Table 14: Average values of total

suspended solids ppm in

Bellal Lake

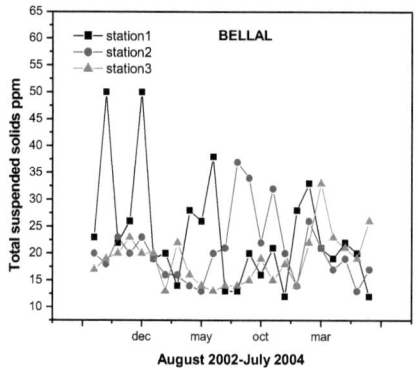

Fig 18: Monthly variations of

Total suspended solids ppm in

Bellal Lake

Zooplanktons

Zooplanktons occupy an intermediate position between the autotrophs and the carnivores and form an important link in aquatic food webs (Trivedy & Goel 1986). Rotifera, Cladocera Ostarcoda , Diptera and Copepoda dominate the zooplanktons in fresh water. Protozoa also forms a significant part of the fresh water zooplankton, especially in systems exhibiting advanced trophic state. The pollution loads affect the presence of zooplankton, their number and behavior. They can be treated as pollution indicators. In the Bellal Lake the following Zooplanktons were found and the results are discussed.

Brachionus, Keratella, Epiphans, Monostyla and Proales representing the Rotifers are observed.

Brachionus

In the present investigation, monthly variations of Brachionus are shown in the Fig. 19. Brachionus species ranged between 2000 org/L to 49000 org/L.

The number of organisms were more at station I and station II in January and minimum in August the average mean values are 19430.6±13434.65 org/L. Brachionus organisms are negatively correlated with temperature and positively correlated with dissolved Oxygen.

Stations	N	Mean	Std. Deviation	Std. Error
1	24	22040	15000.18	3.10
2	24	16400	11200.24	2.31
3	24	20000	14000	2.88
Total	72	19430	13434.65	1.61

Table 15 Average values of

Brachionus org/L in

Bellal Lake

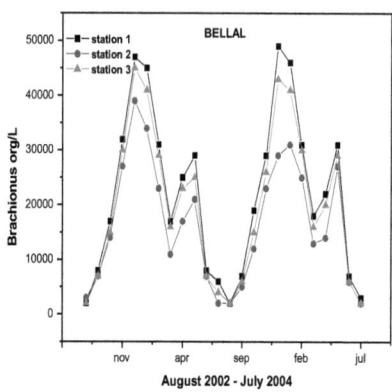

**Fig 19: Monthly variations of Brachionus org/L in
Bellal Lake**

Keratella

In the present investigation, monthly variations of Keratella are shown in the figure 20. Keratella species ranged between 2000 org/L to 28000 org/L

The number of organisms were more at Station I and Station II in the month of December and minimum in August with an average values of 8847.2±6660.8 org/L.

Keratella are positively correlated with dissolved oxygen, biological oxygen demand, total alkalinity, chlorides and phosphates and have no significant correlation with the parameters like P^H, total hardness, calcium, magnesium, and these are not significantly negatively correlated with any other parameters except with the temperature, and total suspended solids. The average values of keratella in the Bellal Lake are given in the Table 16.

Stations	N	Mean	Std. Deviation	Std. Error
1	24	9840.00	7400.5	1.09
2	24	8541.67	6600.8	.88
3	24	8166.66	5900.2	.96
Total	72	8847.22	6660.8	.56

Table 16 Average values of

Keratella org/L in Bellal Lake

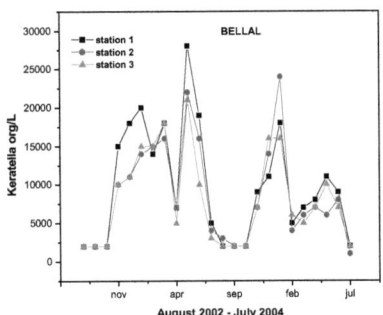

Fig 20: Monthly variations of Keratella org/Lin

Bellal Lake

Epiphanes

In the present investigation, Epiphanes species ranged between 1000 org/L to 20000 org/L with an average of 10000.97±6050 org/L.

The average numbers of epiphanies organisms in the present study are given in the Table.17. The numbers of Epiphanes organisms are maximum at all the stations in December and they were found minimum in the month of August. Epiphans positively correlated with dissolved oxygen, biological oxygen demand, total alkalinity, calcium, magnesium and Nitrates. Monthly variations of Epiphans are shown in the Fig. 21.

stations	N	Mean	Std. Deviation	Std. Error
1	24	11583.75	6736.76	1.44
2	24	9750	5712.30	1.26
3	24	9850.00	5692.6	1.26
Total	72	10000.97	6050	.76

Table 17 Average values of

Epiphans org/L in Bellal Lake

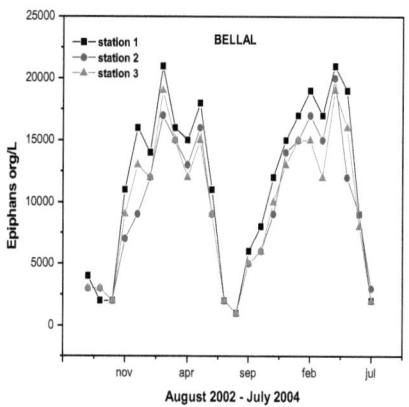

Fig 21: Monthly variations of Epiphanes

org/L in Bellal Lake

Proales

In the present investigation Proales species present ranged between 1600 org/L to 26000 org/L. The numbers of organisms were more at Station I and Station II in January and minimum in October. Average mean values of proales were 8400.00±7000.5 org/L. Monthly variations of Proales are shown in the Fig. 22. Proales are positively correlated with dissolved oxygen, biological oxygen demand, and total alkalinity, whereas negatively correlated with temperature and total suspended solids. The average numbers of Proales in the present study are given in Table.18.

Stations	N	Mean	Std. Deviation	Std.Error
1	24	9500	7800	1.66
2	24	7100	6000.4	1.29
3	24	8560	7250.5	1.55
Total	72	8400.00	7000.5	.86

Table 18 Average values of

Proales org/L in Bellal Lake

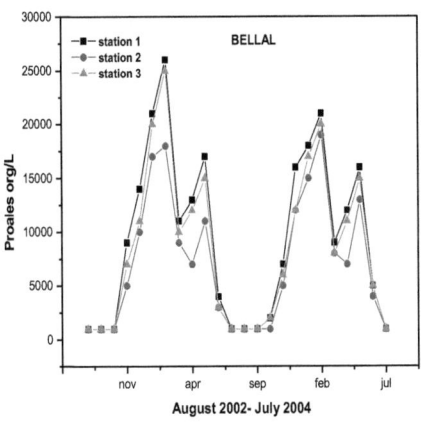

Fig 22: Monthly variations of Proales

org/L in Bellal Lake

Monostyla

In the present study, Monostyla species ranged between 4000 to 42000 org/L with an average mean values 15000.8±10578.59org/L. The no of organisms were more at station I and station III in February and minimum in July. Monthly variations of Monostyla are shown in the Fig.23.

Monostyla is negatively correlated with temperature only and these organisms are positively correlated with the dissolved oxygen and they are not showing any significant correlation with other physico-chemical complexes.

Stations	N	Mean	Std. Deviation	Std. Error
1	24	16000.41	11804.49	2.57
2	24	13666.66	9494.46	2.12
3	24	14666.66	10436.83	2.30
Total	72	15000.8	10578.59	1.34

Table 19 Average values of

Monostyla org/L in Bellal Lake

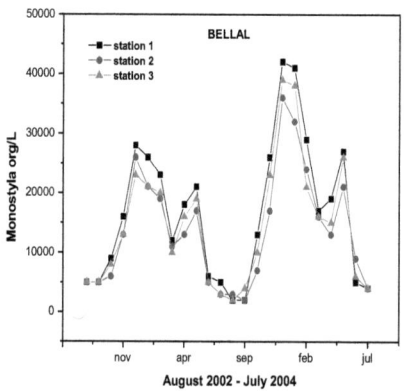

Fig 23: Monthly variations of Monostyla

org/L in Bellal Lake

The seasonal fluctuations of rotifers in the Bellal Lake revealed the highest peak in winter. Presence of rotifers in the form of bloom in winter season could be due to lowering in temperature. These results were in contrast with Padmanabha et al (2006) who studied in the lakes of Mysore.

Cyclops

Cyclops the representatives of copepods occupied second position in the present investigation. Cyclops ranged between 4000 org/L and 45000 org/L with an average of 18555.55±11810.7 org/L in the Bellal Lake.

They are found to be more in the month of February and minimum in the month of August. These observations are supported by Malathi (2002) in Hussain Sagar Lake. The monthly variations are shown in the Fig.24. Cyclops are negatively correlated with temperature, calcium, magnesium and nitrates and positively correlated with dissolved oxygen and total solids. The average mean values of cyclops are represented in Table 20.

Stations	N	Mean	Std. Deviation	Std. Error
1	24	21458.33	14400.0	3.194
2	24	14541.66	7740.2	1.889
3	24	19666.66	13285.0	2.992
Total	72	18555.55	11810.7	1.605

Table 20 Average values of Cyclops org/L

in Bellal Lake

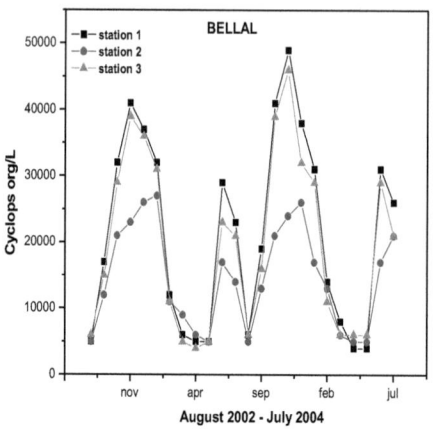

Fig 24 Monthly variations of

Cyclops org/L in Bellal Lake

Nauplius larva

The population density of Nauplius larvae ranged between 1000 org/L to 17000 org/L with an average 4097.2±4503.3 org/L in the Bellal Lake. The average values of nauplius in the present study are given in Table 21. They were maximum in October and minimum in the month of August. The monthly variations are shown in the Fig.25. Nauplius larvae are positively correlated with temperature, biological oxygen demand, calcium, magnesium and nitrates

Stations	N	Mean	Std. Deviation	Std. Error
1	24	4500	5047.5	.4616
2	24	3541.6	3844.5	.2709
3	24	4250	4618	.4094
Total	72	4097.2	4503.3	.2319

Table 21. Average values of

Nauplius org/L in Bellal Lake

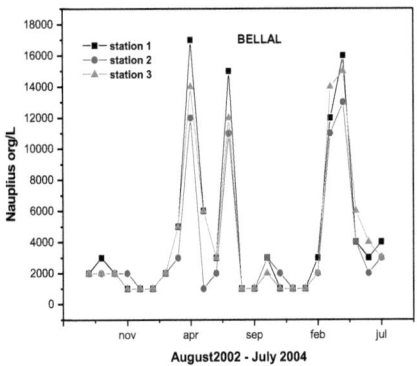

Fig 25. Monthly variations of

Nauplius org/L in Bellal Lake

Daphnia

The daphnia comes under Cladocerans and they have occupied third position in the abundance in Bellal Lake. The average values of daphnia in the present study are given in the Table 22.

Daphnia found maximum number in winter at station I and minimum at station III in the month of July. Similar observations were made by Pawar and Pulle (2005) in Pethwaj dam. Their range is between 2000 to 21000org/L

with an average 9069.4±5431.8 org/L. This range was similar to the findings of Somani and Pejawar (2004) in Masundra lake Thane, Maharashtra. The variations and population density of daphnia is presented in Fig. 26. Daphnia showed negative correlation with temperature. These are positively correlated with the parameters like dissolved oxygen.

Stations	N	Mean	Std. Deviation	Std. Error
1	24	10100.2	6162.0	1.28
2	24	7830.3	5155.5	1.08
3	24	9208.3	4978.0	1.04
Total	72	9069.4	5431.8	.66

Table 22 Average values of

Daphnia org/L in Bellal Lake

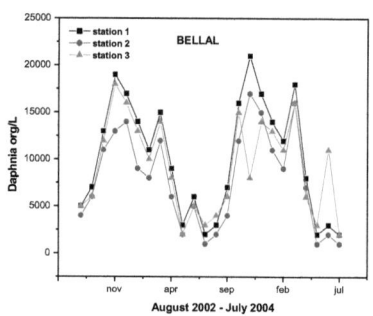

Fig 26 Monthly variations of Daphnia

org/L in Bellal Lake

Cerodaphnia

The Cerodaphnia also comes under cladocerans and were maximum number in winter months i.e. December at all stations and were minimum at station III in rainy months i.e. April, May, June July and August. In the present study, the cerodaphnia ranged between 4000 and 120000 org/L with an average 6810.44±2243.5org/L. Their monthly variations are shown in the Fig.27.

Cerodaphnia organisms are positively correlated with the dissolved oxygen and negatively correlated with temperature. The average values of cerodaphnia in the Bellal Lake are given in the Table 23.

Stations	N	Mean	Std. Deviation	Std. Error
1	24	6958.3	2493	0.50
2	24	6966.6	2057.1	0.41
3	24	6833.3	2180.2	0.44
Total	72	6810.44	2243.5	0.26

Table 23 Average values of

Cerodaphnia org/L in Bellal Lake

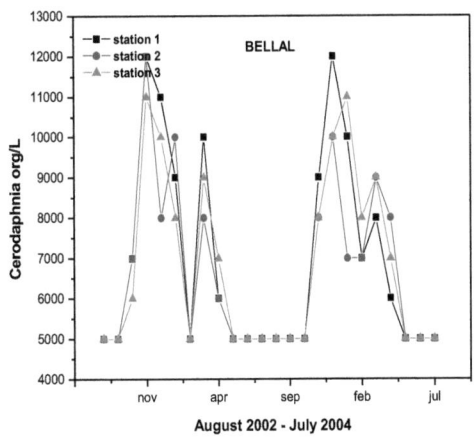

Fig 27 Monthly variations of

Cerodaphnia org/L in Bellal Lake

Cypris

Cypris are included in ostracods they occupy the second position. Their range was 2000 org/L to 17000 org/L with an average 7100+4846.2 org/L. They found maximum in the March at all the stations and less number in the month of August. These results were in conformity with Mishra (2005).

Their monthly variations of cypris organisms are shown in the Fig.28. Cypris organisms are negatively correlated with the dissolved oxygen, biological oxygen demand, total alkalinity, calcium, magnesium, chlorides and nitrates and they are positively correlated with P^H.

Stations	N	Mean	Std. Deviation	Std. Error
1	24	7700	5449.2	1.23
2	24	6450	4343.7	1.01
3	24	7200	4745.5	1.09
Total	72	7100	4846.2	0.63

Table 24 Average values of Cypris org/L in Bellal Lake

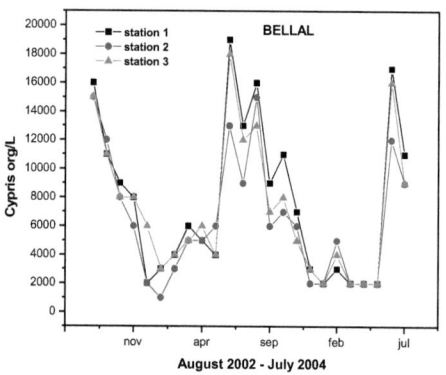

Fig 28: Monthly variations of Cypris org/L in Bellal Lake

Chironomid

Chironomid are the larvae of Diptera. In the present study Chironomid larvae ranged between 5000 org/L and 12000 org/L with an average 6819.4±1909.4 org/L. They were found to be maximum in the month of March at all the stations and less number in the month of August. They were found to be maximum in all the months. The average values are given in Table 25.

Chironomid are negatively correlated with temperature, magnesium and positively correlated with calcium, nitrates and phosphates. The monthly variations are shown in the Fig.29.

Stations	N	Mean	Std. Deviation	Std. Error
1	24	7458.3	2413.3	.49
2	24	6000	1251.0	.25
3	24	7000	2064.1	.42
Total	72	6819.4	1909.5	.24

Table 25. Average values of Chironomid

org/L in Bellal Lake

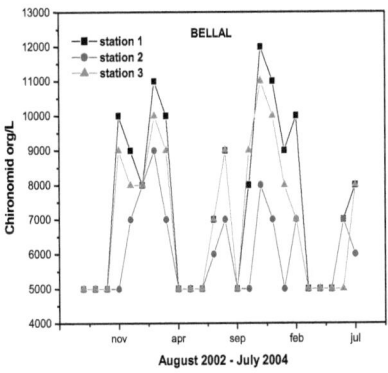

Fig .29. Monthly variations of Chironomid

org/L in Bellal Lake.

Conclusions

Remarkable increase in human population during recent decades has resulted in major changes in the ecology of water bodies. This intensified the use of these water bodies as receiving dumping sites, for disposal of an unprecedented amount of human and agricultural effluents. As a result many water bodies have undergone or are undergoing eutrophic or hypertrophic transformation from oligotrophic state. Such changes are most notable in the water bodies close to populated towns and cities.

Despite increasing awareness and concern for the environment, most of the water bodies are highly polluted and eutrophic. It has not only affected the human use of water (for drinking and recreation) but also for aquatic life. There has been a general decline in capturing fisheries and many plant and animal species disappeared from many water bodies.

At the same time excessive growth of algal blooms and noxious macrophytes are becoming common everywhere. The need for conservation once rooted deep in Indian Culture is now being realized again.

Considerable high quality research supported with advanced instrumentation has been conducted on many Indian lakes by investigators both with in India and from abroad. Nevertheless data and information on these lakes remain relatively sparse in the mainstream literature.

Several characteristics of lakes make them ideal study sites to advance our basic understanding of ecosystem dynamics. Man-made lakes have to receive much attention in India because they are being used for multiple purposes including drinking, irrigation and industrial purposes. These aquatic environments present ecological features that lead to the establishment of a very dynamic system in which the plankton community play an important role.

Human population around the Bellal lake has made a direct and indirect impact on this water body. The agricultural practices based on use of fertilizers, pesticides as well as irrigation and domestic wastes have resulted in degradation in water quality both directly by discharging wastes and indirectly by increased demand for water.

The Bellal lake is characterized by many point and non-point sources of organic and inorganic pollution. A more important non-point source of pollution in Bellal Lake is the agricultural runoff.

On the basis of data collected in present study following conclusions may be drawn:

(1)   Since there are marginal variations in P^H of different sampling sites in the present study it does not affect the fish population and other organisms of the aquatic eco-systems because aquatic life is mainly affected when the P^H is either below 5 or above 11.

(2)   The dissolved oxygen levels in the Bellal Lake are well within the permissible limits.

(3)   Calcium is very important micronutrient influencing both flora and fauna of ecosystem which plays potential role in metabolism and development. In the present investigations amount of calcium is found higher than magnesium.

(4)   Observations indicate that high concentrations and rising trend of total suspended solids is the result of continuous organic input into the lake.

(5)   Values of nitrates and phosphates are suggesting that these nutrients show increasing trend in specific time periods when the lake receives agricultural runoff and domestic waste.

(6)   Data on various physico-chemical characteristics vary for different sites and different months in Bellal lake.

(7)   Variations in zooplankton population may be correlated with physico-chemical characteristics. Maximum population of zooplanktons was recorded in winter months.

(8)   Monthly distribution of zooplankton shows the following pattern in Bellal lake Rotifera >   Copepoda > Cladocera > Ostracoda>Diptera, These planktons are good indicators of water quality of the lake and their ecological significance is more suggesting to control the nutrient load which are key sources to eutrophication process.

(9)   The lake is characterized by high diversity of plankton. Our healthy atmosphere is a heritage to us. We have to use it safely and protect it from all unwanted nuisances, so that, we could handover a healthy protected environment to coming generation.

References

Sharma (1987). Indian Brachiondae, Hydrobiologia 144: 101-105.

Singh (1985). Ecological studies on newly made Tawa reservoir at Raipur Ph.D. thesis Dr.H.SGour University, Sagar

Hussainy, S.U. (1966) studies in the limnology and primary production of a tropical lake. Hydribiol. 30: 335-352.

Smith Raymond C., David Ainley, Karen Baker, Eugene Domack, Steve Emslie, Bill Fraser,James Kennett, Amy Leventer, Ellen Mosley-Thompson, Sharon Stammerjohn, and Maria Vernet (1999): Marine Ecosystem Sensitivityto Climate Change: BioScience Vol. 49 No. 5

Lewis, W.M.Jr. (1987): tropical limnology annual Ann.Rev.Ecol.Syst.8: 159-184.

Gunter Gunkel. (2000). Limnology of an equatorial High Mountain Lake in Ecuador,Lago San Pablo. Limnologica ,30:113-120.

Burton T. (1997) Le plus grand lac du Mexique peut-il être sau, vé? Ecodécision 23, 68–71.

Frenchel T., (1987). Ecology of Protozoa. The Biology of free living phagotrophic protests. Science tech publishers, Madison (197p.)

Laybourn-Parry,J., (1992). Protozoan Plankton Ecology. Chapman& Hall,London 231p.

Porter,K.G., Sherr,E.B.,Sherr E.F. Pace M.,.Sanders R.W. (1985) Protozoa in planktonic foodwebs J.Protozool 32,409-415.

Welch, P. S. (1952). Limnology 2nd Ed. McGraw Hill. London.

Forel, F.A. (1892-1904). Le Leman. Monographie limnoloique. 3 vols. Lausanne

Chumly (1910): Bathymetrical survey of the Scottish fresh-water lochs (Volume 2) Oxford publications

Zacharias.O (1891) Die Tier- und Pflanzenwelt des Süsswassers. J.J.Weber Verlags buchhandlung, Leipzig, p 749

Agassia (1850): Lake Superior: its Physical chatacter,Vegetation and Animals,compared with those of other and similar regions Bibliographia Zoologiae et Geologiae Boston: Gould, Kendall and Lincoln

Stimson (1870). The gorgonians of the Brazilian coast. Jour. Acad. Nat. Sci. Philadelphia, 2(15): 373-404.

Smith and Verril. (1871). Report upon the invertebrate animals of Vineyard Sound and adjacent waters, with an account of the physical features of the region. U.S. Comm. Fish Fish., Rep. Comm.:295-778, pl. 1-38

Green, J. D. (1972): Studies on the zooplankton of Lake Pupuke. Tane *13:* 77-98.

Jolly, V. H. (1965): Diurnal surface concentration of zooplankton in Lake Taupo, New Zealand. Hydrobiologia 25: 466-72.

Bagchi S. N. (1986) Cyanobacterial antibiotic cyanobacterin and its role in biological control of algal blooms. Annual Conf. Of MP Council of Sci. & Tech.

Green, J.D, Chapman, M. A.; Jolly, V. H. (1974): The limnology of a New Zealand reservoir, with particular reference to the life histories of the Copepoda *Boeckella propinqua* Sars and *Mesocyclops leuckarti* Claus. Internationale Revue der gesamten Hydrobiologie 59: 441-87.

Prasshad, H.N. (1916). The seasonal conditions governing the pond life in the Punjab. J. Assist .Soc.Bengal-12:142-145.

Sewell.,R.B.J (1924). Observations on growth in certain molluscs and on changes connected.... Fauna of British India

Sewell.,R.B.J.(1934). Studies on the biomass of fresh water in India. Int.Rev.Ges.Hydrobiol-Hydrograph, 31,203-208.

Pruthi. H.S. (1933) studies on bionomics of fresh water in india, seasonal changes in physical and chemical conditions of the water of the tank in the Indian museum campus, international reviews geo.Hydrobiologia 28,46-47.

Chopra B (1930) First record of occurrence in India of the ancient sub-order Phreatoicoidea (Crustacea: Isopoda). Proceedings of the Indian Science Congress 34: 176.

Ganapathi, S.V. and Pathak, C.H. (1959). Primary productivity in the Sayaji Sarovar (Amman, Made lake) at Baroda. Proc. Semi. Eco. and Fish. Fresh water reservoirs, 25-37.

Vasisht, H.S and Sharma B.K, (1976). Seasonal abundance of rotifer population in a fresh water pond in Ambala city (Haryana) India. J.Zool.Soc.India, 28:35-44.

Aziz Hussain.M (1977). Ecobiology of fresh water Protozoa. PhD thesis Osmania University, Hyderabad

Krishna kumar (1976): observations on diurnal variations in hydrobiological conditions of two fish ponds, Hyderabad, India .comp. physiol.ecol 3(3):111-114.

I.S.Rao (1982): Ecobiology of Manjeera Reservoir. Ph.D thesis Osmania University,Hyderabad.

Jaya Devi , M., (1985) Ecological studies of the Limnoplankton of three fresh water bodies of Hyderabad Ph.D thesis Osmania University, Hyderabad.

Tandon K.K. and Singh. (1988) River Pollution and fish- A report on pollution calamity of Uttarpradesh. In: Ecology and pollution of Indian Rivers (R.K. Trivedi ed.) Ashish Publishing House, New Delhi, India.

Nayar, C.K.G., (1970). Studies on the rotifer population of two ponds at Pilani, Rajasthan. J.Zool.Soc.India, 22:168-185.

Mishra G.P, Yadav, A.K. (1981) Zooplankton and their seasonal abundance in a sewage collecting river at Gwalior Madhya Pradesh, Advances in Limnology 1-45.

Gopal B (1990): Ecology and sustainable development national institute of ecology India publications.

Kaul and Raina (1982): Limnological studies of two ponds in Jammu, Physicochemicalparameters.J.Environ.Biol.10:134-144

ZutshiD. P.,B. A. Subla,M. A. KhanandAshwani Wanganeo(1980)Comparative limnology of nine lakes of Jammu and Kashmir Himalayas Hydrobiologia 101– 112.

Zutshi, D.P. and. Khan A.U. (1988). Eutrophication gradient in Dal Lake, Kashmir. Indian J. Environ. Hlth., 30(4): 348-354.

Das A.K. (2000) Limno-chemistry of some Andhra Pradesh reservoirs J.Inland Fish.Soc.India 32(2)37-44

Panigrahy, R.C. (2000). The Chilka Lake, A sensitive coastal Ecosystem of Orrisa, east Coast of Indian Ocean. J. of Indian Ocean Studies, 7(2,3): 223-242.

Gupta, S.C., Rathore, G.S. and G.C.D. Mathura (2001). Hydrochemistry of Udaipur Lakes. Indian J. Environ. Hlth., 43(1): 38-44.

Seshavathranam, V. (1992). Ecological Status of Kolleru Lake: A review. Wetlands of India (K.J.S. Chathran, Editor). 172-200.

Venkata Ramanaiah Solanki, Md Hussain; and S.Sabita Raja (2010) Water quality Assessment of Lake Pandu Bodhan, Andhra Pradesh State, India. Environ Monit. Assess. 163: 411–419.

George M.G. (1968) Case construction,movement, spatial distribution and substrate selection in the larvae of *Chironomus riparius*. Journal of Experimental Biology, 50, 247–253.

Moitra, S.K. and Battacharya (1965): seasonal cycle of rotifers in fresh water fish pond in Kalyani West Bengal. Proc. Symp.Rec.Adv.Prop.Ecol 359-361.

Khan A.A. (1969) Diurnal variations in the pond Moat at Aligarh. J.Inland.Fish.Soc.India: 146-156.

Shah .H. K.L.sehagal ,Mitra and N.C.Nandi (1971) studies on seasonal and diurnal variations in physico-chemical and biological conditions of a perennial tank J.Inland.Fish.Soc.3:80-102.

Jana.B.B. (1973). Seasonal periodicity of planktons in a fresh water pond in West Bengal. Germaten Hydrobiol 58(1): 127-143.

Tiwari R.D. (1977) ecological problems in the upper lake, Bhopal- problems and Suggestions Eco.Env. Conserv, 5(2), 115-117

Krishna murthy (1988) hydrobiological studies in Gandhi sagar lake seasonal variations in plankton .hydrobiol.25. (3): 501-514

Sreenivasan, A. (1969). An experimental study for the determination of frictional resistance on a model boat under different mud concentrations. Mahasagar—Bull. natn. Inst. Oceanogr. 7 (3 & 4):183-188.

Michael and Victori (1973) Ecology of Shallow and Deep Water Populations of *Pontoporeia hoyi* (Smith) (Amphipoda) in Lake Michigan Freshwater Invertebrate Biology, Vol. 3, No. 3 pp. 118-138

Jyotj. M.A. and Sehgal .H. (1979). Ecology of rotifers of Surinsar, a subtropical fresh water lake in Jammu(J&K) India.Hydrobiol.65(1):25-32.

Rao, K.V. A.K.Khnderkar and D.Vaidyanathan. (1976). Uptake of fluoride by water hyacinth, Eichornia crispers. Indian J.Exp.Biol.11:68-69

Das, Gupta, (1982). Discourse on aquatic polymycetes of India, Indian phytopath, 35:193-216.

Biswas(1964)Fivespeciesofdaphnidae(Crustaceae:cladocera) from Simla Hills in India , with new record of Alona costate sars from kameng Division NEFA. Jour.Zool.Soc.India. 16(1&2):92-98

Jhingran V.G., Natarajan A.V., Banerjea S.M., David A., (1969). Methodology on reservoir fisheries investigations in India. Bull. Cent. Inl. Fish. Res. Inst., Barrackpore.

Manja, S.K. and Kaul, R.K. (1992). Efficacy of a simple test for bacteriological quality of water. Journal IAEM,19:18-20.

Pelczar, M.J., Chan, E.C.S., Krieg, N.R. (1988). Heavy metals and their compounds: Microbiology, 5th ed., McGraw-Hill Book Company, New York, pp.497-498.

Wright I.A.(1986) measuring the impact of sewage effluent on the macro invertebrate community of upland streams. Australian jou.ecol 20(1):142-149.

Pathak, S.P., Bhattacharjee, J.W., Karla, N., and Chandra, S. (1988). Seasonal distribution of Aeromonas hydrophila in river water and isolation from river fish.J.Appl.Bact.65: 347-352.

Abbasi, S.A., Bhaliya, K.S., Kunhi, A.V.M., and Soni, R S. (1996). Studies on the limnology of Kuttadi Lake (North Kerala), Ecol. Env. & Cons.2: 17-27.

Shrivastava, R.K., Shrivastava Seema and Shukla, A.K. (2001). River pollution in India: A brief Review. J.Environ.Research, 2(2): 111-115.

Kant, S. and Kachroo, P. (1975). Limnological studies in Kashmir lakes. II. Diurnal movement of phytoplankton. J. Ind.Bot.Soc.54 (1/2): 9-12.

Zutshi,D.P. and Vaas K.K.(1977). Limnological studies on Dal Lake chemical factors,Indian J.Ecolo 5(1):90-97.

Vaas K.K. (1982) on the trophic status and conservation of Kashmir lakes.Hydrobiologia 68(1):9-15.

Ganapathi, S.V. (1943). The limnology of two minor irrigation reservoirs near Madras. 1.enalupam reservoir, Hydrobio.8 (3-4): 365-438.

Zafar, A. R. (1968). Certain aspects of Distribution Pattern of Phytoplankton in the Lakes of Hyderabad. In symposium on Recent Advances in Tropical Ecology Internat. Soc. Trop. Ecol. Varanasi-5, India.

Munawar, M. (1970). Limnological studies on fresh water ponds of Hyderabad India. Hydrobiology 31: 101-128.

Rao, V. S. (1975). An Ecological Study of Three Fresh Water Ponds of Hyderabad-India, III. The phytoplankton (Volvocales, Chlorococcales, Desmids), Hydrobiologia, vol. 47 (2): 319–337.

Chourasia Sadhana (1985) seasonal fluctuations as of zooplankton in Burha Tank water Raipur India J.Env.Prof.16 (2):140-142.

Hegde and Kale (1985). Ecological studies in ponds and lakes of Dharwad.trophic status Phykos. 23(1) 71-74.

Rao. A.M., (2004). The studies on fisheries in a Tilapia dominated reservoir Ph.D Thesis. Osmania University, Hyderabad.

Shanty, K, Rama swamy, K and Lakshmana Perumala Swamy, P. (2002). Hydro biological study of Singanallur Lake at Coimbatore, India. Nature Environment and Pollution Technology. 1(2): 97-101.

Eswarlal sedamkar and S.B.angadi (2003). physico-chemical parameters of two fresh waterbodies of gulbarga-india, with special reference to phytoplankton. Poll Res.22(3):411-422.

Usha Awasthi and Sunil Tiwari (2004) seasonal trends in abiotic factors lentic habitat. Eco. Env & cons. 10(2): 165-170.

Savita dixit, S.K. Gupta and Suchi Tiwari (2005). Nutrient overloading in fresh water lake in Bhopal, India. Electronic green journal 21.

Shiva kumar and K.Altaf (2005) Biodiversity of fresh water zooplankton in and around Chennai UGC National Seminar on Biodiversity studies on Zooplankton

Vijay Kumar and Murali Jadesh (2005) effect of physico-chemical factors on the seasonal abundance of Rotifera in Kagina Reservoir. Eco. Env & cons. 4(2): 135-140

Kudari V.A., R.D. Kanamadi and G.G. Kadavaram (2006). Limnological studies of atteveri and Bachanki reservoirs of UttarKannada District Karnataka India. Eco, Env. &Cons.12 (1); 87-92.

Ravi kumar.M   Manjappa, S. Kiran, B.R. and Puttaiah, E.T. (2006) phyto plankton periodicity in relation to abiotic factors in Kulahalli tank near Harapanahalli, Karnataka. Nature Environment and Pollution Technology pp. 157-161

Bonner, L. A., Diehl, W. J. & Altig, R. (1997): Physical, chemical and biological dynamics of five temporary dystrophic forest pools in central Mississippi. Hydrobiologia 353: 77-89.

Williums, W. D. (1985): Biotic adaptations in temporary lentic waters, with special reference to those in semi-arid and arid regions. Hydrobiologia 125: 85-110.

American Public Health Association (APHA). (1989) Standard Methods for the Examination of Water and Wastewater, 17th edn. American Public Health Association, Washington DC.

Trivedy, R.K.and Goel, P.K.(1986). Chemical and biological methods for water pollution studies. Environmental Publications. Karad.

Hussian, J., Hussain, I., Sharma, K.C. and Arif, M.(2004). Impact of domestic wastewater on Gandhisagar lake Bhilwara, Rajasthan, India. Ecol. Env. & Cons.10 (3): pp.363-367.

Sreenivasan, A. 1974. Limnological features of tropical impoundment Bhavanisagar reservoir (Tamil Nadu), India. Int. Revue. Ges. Hydrobiol. 59 : 327-342.

Chatterjee, C. and Raziuddin. M (2002). Determination of water quality index WQI of a degraded river in Asansol industrial area (W.B). Nature Environmnt and Pollution Technology, 1(2) 181-189.

Ahmad Masood and Krishnamurthy, R. (1990). Hydrobiological studies of Wohar reservoir, Aurangabad (Maharashtra state). Indian. J.Environ.Biol. 11(3): 335-343.

Hussain, J., Hussain, I. and Arif, M. 2004. Characterization of textile wastewater. J. Indl. Pol. Cont. 20(1): 137-144.

Devaraju, T.M, Venkatesha,M.G. and Surendra singh (2005). Studies on the physico-chemical parameters of Maddur Lake with reference to suitability for Aquaculture Nature Environment and Pollution Technology, 4(2) 287-290.

Shastri, Y. and Pendse, D. S. 2001. Hydrobiological study of Dehikhuta reservoir. J. Environ. Biol. 22: 67-70

Prithwiraj Jha and Sudip Bharat (2003) Hydrobiological study of lake Mirik in Darjeeling Himalayas.J.Environ.Biol.24 (3), 339-344.

Peter, K. J., (1974). An appraisal of the larval fishes of the Arabian Sea. Seafood Export J 6(6): 47-49

Bagde V.S. and A.K. Varma (1985) physico-chemical characterstics of JNU lake at new Delhi. Indian J.Ecol 12(1) 151-156.

ICMR Manual of standards of quality fordrinking water supplies. ICMR, New Delhi.(1975).

Lehr, J. H., Gass, T.E. Pettyjohn, W.A. and DeMarre, J. (1980). Domestic water treatment McGraw-Hill Book Co.

Salodia, P.K. (1996) fresh water biology. Surabhi Publications.

Singh, Y., (1979). Ecological studies on the river algae with special reference to Bacillariophyceae. Ph.D. Thesis Univ.Luck., Lucknow, India

Thilaga,A, S. Subhashini, S. Sobhana and K. Logan Kumar (2005) Studies on nutrient content of the ooty lake with reference to pollution Nature Environment and Pollution Technology,) 4(2):47-52.

Mazumdar. P (2000): "Rural Drinking Water: Some Aspects", Proceedings of West Bengal State Science & Technology Congress,. SOC-Env. 47

Padmanabha B. and S.C. Belgali (2006) comparative study on population dynamics of rotifers and water quality index in the lakes of Mysore. Nature Environment and Pollution Technology, 5(1) 107-109.

Malathi D (2002) Ecological studies of fresh water zooplankton in hussain sagar lake. Ph.D. thesis osmania university, Hyderabad.

Pawar, S.K., V.R. Madlapure, J.S. Pulle (2003). Study of Zooplanktonic Community of Sirur Dam water near Mukhed in Nanded Dist (M.S.) India, J.of Aqua.Biol.18 (2), 37-40.

Gonzalves, E. A. & Joshi, D. B. (1946). Freshwater algae near Bombay. I. The seasonal succession of the algae in a tank at Bandra. J. Bombay Nat. Hist. Phil. Soc., 46: 154–176.

Somani and Pejawar (2004) crustacean zooplankton of lake masundra Thane Maharashtra J.Aqua.Biol.19(1):57-60.

Mishra N.K. (2005) Limnological study and plankton,Interim fish culture of Gurma Bundh Rewa District (M.P.) Thesis submitted A.P.S. University (M.P.)